THE NEW GROVE
BEETHOVEN

THE NEW GROVE
DICTIONARY OF MUSIC AND MUSICIANS
Editor: Stanley Sadie

The Composer Biography Series

BACH FAMILY

BEETHOVEN

HANDEL

HAYDN

MASTERS OF ITALIAN OPERA

MOZART

SCHUBERT

SECOND VIENNESE SCHOOL

THE NEW GROVE

BEETHOVEN

**Joseph Kerman
Alan Tyson**

**PAPERMAC
MACMILLAN LONDON**

First published in
The New Grove Dictionary of Music and Musicians,
edited by Stanley Sadie, 1980

First published in paperback with additions 1983 by
PAPERMAC
a division of Macmillan Publishers Limited
London and Basingstoke

Associated companies in Auckland, Dallas,
Delhi, Dublin, Hong Kong, Johannesburg,
Lagos, Manzini, Melbourne, Nairobi,
New York, Singapore, Tokyo, Washington
and Zaria

ISBN 0 333 35385 4

Filmset in Times by Filmtype Services Limited,
Scarborough, North Yorkshire

Printed in Hong Kong

Contents

List of illustrations

We are grateful to the following for permission to reproduce illustrative
material: Deutsche Staatsbibliothek, Berlin (figs.1, 3); Beethovenhaus,
Bonn (cover, figs.2, 4, 9, 12); British Library, London (figs.5, 8);
Staatsbibliothek Preussischer Kulturbesitz, Musikabteilung, Berlin
(figs.6, 11); Historisches Museum der Stadt Wien (fig.7); Gesellschaft
der Musikfreunde, Vienna (fig.10); Museum der Bildenden Künste,
Leipzig (fig.13).

General Abbreviations

A	alto, contralto	ob	oboe
acc.	accompaniment, accompanied by	orch	orchestra, orchestral
		org	organ
arr.	arrangement, arranged by/for	ov.	overture
		pt.	part
B	bass [voice]	pubd	published
b	bass [instrument]		
bn	bassoon	qnt	quintet
		qt	quartet
cl	clarinet		
conc.	concerto	*R*	photographic reprint
db	double bass	recit	recitative
ded.	dedication, dedicated to	red.	reduction, reduced for
		rev.	revision, revised by/for
edn.	edition		
eng hn	english horn	S	soprano
		str	string(s)
facs.	facsimile	sym.	symphony, symphonic
fl	flute		
frag.	fragment	T	tenor
hn	horn	transcr.	transcription, transcribed by/for
inc.	incomplete		
		trbn	trombone
Jb	Jahrbuch [yearbook]		
kbd	keyboard	v, vv	voice, voices
		va	viola
mand	mandolin	vc	cello
movt	movement	vn	violin

Symbols for the library sources of works, printed in *italic*, correspond to those used in *RISM*, Sec. A.

Bibliographical Abbreviations

AcM	*Acta Musicologica*
AMZ	*Allgemeine musikalische Zeitung*
BeJ	*Beethoven-Jahrbuch*
BMw	*Beiträge zur Musikwissenschaft*
FAM	*Fontes artis musicae*
GfMKB	*Gesellschaft für Musikforschung Kongressbericht*
Grove 1	G. Grove, ed.: *A Dictionary of Music and Musicians*
JAMS	*Journal of the American Musicological Society*
Mf	*Die Musikforschung*
MGG	*Die Musik in Geschichte und Gegenwart*
ML	*Music and Letters*
MMg	*Monatshefte für Musikgeschichte*
MMR	*The Monthly Musical Record*
MQ	*The Musical Quarterly*
MR	*The Musical Review*
MT	*The Musical Times*
NBJb	*Neues Beethoven-Jahrbuch*
NRMI	*Nuova rivista musicale italiana*
NZM	*Neue Zeitschrift für Musik*
ÖMz	*Österreichische Musikzeitschrift*
PNZ	*Perspectives of New Music*
PRMA	*Proceedings of the Royal Musical Association*
RMI	*Rivista musicale italiana*
SMw	*Studien zur Musikwissenschaft*
SMz	*Schweizerische Musikzeitung/Revue musicale suisse*
ZfM	*Zeitschrift für Musik*
ZmW	*Zeitschrift für Musikwissenschaft*

Preface

This volume is one of a series of short biographies derived from *The New Grove Dictionary of Music and Musicians* (London, 1980). In its original form, the text was written in the mid-1970s, and finalized at the end of that decade. For this reprint, the text has been re-read and modified by the original authors and corrections and changes have been made. The work-list (originally compiled by Douglas Johnson) and bibliography (originally compiled by William Drabkin) have been brought up to date and incorporate the findings of recent research.

The fact that the texts of the books in the series originated as dictionary articles inevitably gives them a character somewhat different from that of books conceived as such. They are designed, first of all, to accommodate a very great deal of information in a manner that makes reference quick and easy. Their first concern is with fact rather than opinion, and this leads to a larger than usual proportion of the texts being devoted to biography than to critical discussion. The nature of a reference work gives it a particular obligation to convey received knowledge and to treat of composers' lives and works in an encyclopedic fashion, with proper acknowledgment of sources and due care to reflect different standpoints, rather than to embody imaginative or speculative writing about a composer's character or his music. It is hoped that the comprehensive work-lists and extended bibliographies, indicative of the origins of the books in a reference work, will be valuable to the reader who is eager for full and accurate reference information and who may not have ready access to *The New Grove Dictionary* or who may prefer to have it in this more compact form.

S.S.

*

The article on Beethoven for the original edition (1879–89) of Grove's *Dictionary of Music and Musicians* was written by the editor himself. Reprinted in the second, third and fourth editions, Grove's contribution was replaced only in the fifth edition (1954). On its retirement it was published as a book, together with similarly cashiered articles by Grove on Schubert and Mendelssohn. This was appropriate enough. For the real reason for the articles being withdrawn was not that they were too long or out of date – though that was certainly the case – but because they were too much like little books. They were too personal for the standards deemed suitable in the 1950s for an authoritative and objective great dictionary of music.

Today we are likely to agree that a good dictionary article about a composer should fulfil a different function from a book about him, or an extended essay of the 'life and works' category. If Grove thought differently, that is a reflection of his personal attitudes and those of his era. A book on Beethoven will be expected to provide a forceful interpretation; whether it is as slight as J. W. N. Sullivan's or as multivoluminous as Romain Rolland's, the picture of Beethoven as man and artist that it presents will be an individual one, painted by the author with his own colours, his own brush or palette knife. This is no doubt what a reader will be looking for in a book on

Beethoven – in any case, this is what he will find. And he will find the author's name on the title-page along with Beethoven's.

The reader of a dictionary article will probably not be looking for an interpretation but for uncontroversial data. (He may or may not notice the author's name tucked away at the end of the article.) A monument like *Grove*, designed to be an authoritative reference work, or as near as one can come to that ideal, will be turned to by students, historians and journalists, and should appropriately emphasize factual elements rather than evaluative ones. It is, after all, no accident that the work-lists and bibliographies of *The New Grove* were prepared with such care and have been received with such appreciation.

This distinction between a book and a dictionary article was one that the present authors had in mind while writing the *Beethoven* entry; and we feel impelled to remind our readers of it, now that the article is appearing in the form of a book. It will help to explain certain characteristics of the article, in its dimensions as well as in its tone. To mention what is perhaps the most obvious: the larger space devoted to Beethoven's life than to his work does not reflect any view of absolute priorities, only a view of proprieties for the writing of dictionary articles. For when people consult a dictionary they are more likely to be seeking facts (which for the life can be established with relative certainty) than critical judgments (which are more uncertain, more subjective and more fluid). In regard to the music, we felt a certain constraint against straying too far from what might be regarded as received critical opinion; it was not our job here to build the kind of interpretation that would be expected in a real 'life and works'. To what extent the life, too, should be handled in a bold interpretative manner is certainly an issue that biographers today are required to face. But for a dictionary, once again, we felt that a reasonably complete chronological narrative of the traditional kind would best serve the needs of its readers.

On the occasion of the present republication, a few additions have been made to the sections on Beethoven's music and a few minor errors or oversights have been corrected. A thorough revision, however, was no part of the publisher's scheme and could not have been accomplished so simply. For in fact the decade that has passed between the writing of the article and its publication in the present format has seen a great deal of research into and re-interpretation of Beethoven's life, his music and the interrelations between life and music. We have been able to glance at some of this but not really to absorb it into an older piece of writing. Like Grove's original article, then, and like the one by William McNaught which replaced it in the fifth edition, ours must be read as a product of its era: in the sense both of what was known and thought at the time, and also of the attitudes generally held (and promulgated by the team of editors at *The New Grove*) about what constitutes a functional dictionary entry.

Chapters Two and Four were drafted by the first of the signatories below, Chapters One and Three by the second.

Joseph Kerman
Alan Tyson

CHAPTER ONE

Life

I Family background and childhood

Three generations of the Beethoven family found employment as musicians at the court of the Electorate of Cologne, which had its seat at Bonn. The composer's grandfather, Ludwig (Louis) van Beethoven (1712–73), the son of an enterprising burgher of Mechelen (Belgium), was a trained musician with a fine bass voice, and after positions at Mechelen, Louvain and Liège accepted in 1733 an appointment as bass in the electoral chapel at Bonn. In 1761 he was appointed Kapellmeister, a position which – although he seems not to have been a composer, unlike other occupants of such a post – carried with it the responsibility of supervising the musical establishment of the court.

By his wife Maria Josepha Poll, whom he married in 1733, and who later took to drink, he had only one child that survived. Johann van Beethoven (c1740–1792) was a lesser man than his father. He, too, entered the elector's service, first as a boy soprano in 1752, and continuing after adolescence as a tenor. He was also proficient enough on the piano and the violin to be able to supplement his income by giving lessons on those instruments as well as in singing. In November 1767 he married Maria Magdalena (1746–87), daughter of Heinrich Keverich, 'overseer of cooking' at the electoral summer palace of Ehrenbreitstein, and already the widow of Johann Leym, valet to the Elector of Trier; she was not yet 21. The couple took lodgings in Bonn at 515 Bonngasse. Their first child Ludwig Maria (baptized 2 April 1769) lived only six days; their second, also called Ludwig and the subject of this narrative, was baptized on 17 December 1770. Of five children subsequently born to the couple only two survived infancy: Caspar Anton Carl (baptized 8 April 1774) and Nikolaus Johann (baptized 2 October 1776). Both brothers were to play important parts in Beethoven's life.

Inevitably the early years of the son of an obscure musician in a small provincial town are themselves sunk in obscurity, and though speculation and myth-making have both been productive, facts are rather scarce. It is clear that at a very early age he received instruction from his father on the piano and the violin. Tradition adds that the child, made to stand at the keyboard, was often in tears. Beethoven's first appearance in public was at a concert given with another of his father's pupils (a contralto) on 26 March 1778, at which (according to the advertisement) he played 'various clavier concertos and trios'. A little later, when he was eight, his father is said to have sent him to the old court organist van den Eeden, from whom he may have received some grounding in music theory as well as keyboard instruction. He appears also to have had piano lessons from one Tobias Pfeifer, who lodged for a while with the family, and informal tuition from several local organists. A relative, Franz Rovantini, gave the boy lessons on the violin and viola. His general education was not continued beyond the elementary school, but this was in accordance with the usual custom in Bonn at that time, only a few children going on to a Gymnasium (high school). The comparative brevity of Beethoven's formal education, combined with the fact that most of his out-of-school hours must have been devoted to music, explains some of the gaps in his academic equipment, such as his blindness to orthography and punctuation and his inability to carry out the simplest multiplication sum.

In 1779 a musician arrived in Bonn who was to be Beethoven's first important teacher. This was Christian Gottlob Neefe, who came as the musical director of a theatrical company that the elector took into his establishment. The point at which he began instructing Beethoven is not known. But in February 1781 Neefe succeeded to the post of court organist, a position that evidently required an assistant, and by June 1782, when Neefe left Bonn for a short period, Beethoven was acting as deputy in his absence; he was then $11\frac{1}{2}$. Neefe's estimate of his pupil is contained in a communication to Cramer's *Magazin der Musik* dated 2 March 1783 – the first printed notice of Beethoven:

Louis van Beethoven, son of the tenor singer already mentioned, a boy of 11 years and of most promising talent. He plays the piano very skilfully and with power, reads at sight very well, and I need say no more than that the chief piece he plays is *Das wohltemperirte Clavier* of Sebastian Bach, which Herr Neefe put into his hands ... So far as his other duties permitted, Herr Neefe has also given him instruction in thoroughbass. He is now training him in composition and for his encouragement has had nine variations for the piano, written by him on a march [by Ernst Christoph Dressler], engraved at Mannheim. This youthful genius is deserving of help to enable him to travel. He would surely become a second Wolfgang Amadeus Mozart if he were to continue as he has begun.

3

The reference to Mozart was presumably to the child prodigy and not to the mature composer whose years of fame in Vienna were yet to come; but Neefe's affection for his young pupil and confidence in his ability are plain. The variations on Dressler's march (WoO 63), published by Götz of Mannheim, were Beethoven's first published work.

Further experience came to Beethoven via Neefe in 1783 when his teacher, overburdened with the work of the temporarily absent Kapellmeister Lucchesi, employed him as 'cembalist in the orchestra', not only a position of some responsibility but also one that will have enabled him to hear all the popular operas of the day. The autumn saw the publication of his first significant composition, the three piano sonatas dedicated to the Elector Maximilian Friedrich (WoO 47). Since Beethoven, for all his work on Neefe's behalf, still had no salary, he petitioned the court for an official position as assistant to the court organist; this was granted, but the elector died before the salary, if any, could be fixed.

II Youth

The new elector, Maximilian Franz, brother of the Habsburg Emperor Joseph II, instituted economies on his accession in 1784 that temporarily affected Neefe, some of whose salary was transferred to his pupil. Beethoven's salary as organist was fixed at 150 florins; his duties seem to have been light, leaving him time for composition as well as for general musical development. From 1785, at any rate, dates a set of three piano quartets (WoO 36), possibly intended for dedication to

4

the new elector but not published until after Beethoven's death. About this time, too, he seems to have had violin lessons from Franz Ries, a good friend of the family, and to have begun giving piano lessons himself. In 1787 he spread his wings more widely. Neefe, as quoted above, had declared that the young genius should be given the chance to travel, and in the spring of that year Beethoven visited Vienna. In the absence of documents much remains uncertain about the precise aims of the journey and the extent to which they were realized; but there seems little doubt that he met Mozart and perhaps had a few lessons from him. It seems equally clear that he did not remain in Vienna for longer than about two weeks. The news of his mother's deteriorating health precipitated his sudden journey back. He returned to Bonn to find his mother dying of tuberculosis, and his first surviving letter, to a member of a family in Augsburg that had befriended him on his way, describes the melancholy events of that summer and hints at his own ill-health, depression and lack of financial resources.

For the fortunes of the Beethoven family were in decline. This is perhaps the place for a word or two about Beethoven's parents. The personality of the mother whom he now mourned (she had died on 17 July 1787) does not emerge in very distinctive terms; the accounts speak in conventional phrases of her piety, gentleness and kindness, and of her gravity of manner. This is contrasted, again somewhat conventionally, with Johann van Beethoven's harsher and perhaps even violent temperament. In these years the talents on which he relied to support his family, at no

5

time outstanding, seem to have been observed to decline. An official report of 1784 described his voice as 'very stale', and for some time before his wife's death he had begun to drink heavily, as his mother had done. In 1789, therefore, Beethoven – who was not yet 19 – took the unprecedented step of placing himself at the head of the family by petitioning for half his father's salary to enable him to support his brothers; this was granted, and the old tenor's services were dispensed with. The psychological significance of this act of self-assertion has not escaped his biographers.

The next four years, the last that Beethoven spent in Bonn, can be portrayed in a sunnier light. From 1789, when the musical life of the town under the new elector was fully resumed, Beethoven played the viola in the orchestras both of the court chapel and of the theatre, alongside such fine musicians as Franz Ries and Andreas Romberg (violins), Bernhard Romberg (cello), Nikolaus Simrock (horn) and Antoine Reicha (flute); some of these were to remain almost lifelong friends. This orchestral experience forms a necessary background to the most impressive composition of the Bonn years, the cantata on the death of the Emperor Joseph II (WoO 87).

Joseph II was not merely the elector's elder brother but a powerful symbol of those intellectual, social and political ideas of the 18th century known as the Enlightenment (*Aufklärung*). His reformist ideas found a ready welcome in Bonn among Beethoven's contemporaries and immediate superiors in age, so that the grief caused by the emperor's death in Vienna on 20 February 1790 was no doubt more than merely formal.

On hearing the news four days later the literary society (Lesegesellschaft) of Bonn at once planned a memorial celebration for 19 March. Beethoven was commissioned to produce a cantata, but for unknown reasons the work was not performed. It may be that there was insufficient time to rehearse it; that it was found unimpressive seems unlikely, since in the autumn a second cantata 'On the Elevation of Leopold II to the Imperial Dignity' (WoO 88) was commissioned and completed – though that too seems not to have reached performance.

One further commission was undertaken to please Beethoven's talented and powerful friend Count Ferdinand Waldstein: on 6 March 1791 the count produced a ballet in old German costume, performed by the local nobility, and the music for this *Ritterballett* (WoO 1) was by Beethoven, though his name was not made public. The dedication to the Countess von Hatzfeld of 24 variations for piano on the theme of Righini's arietta 'Venni amore' (WoO 65), published in the summer of 1791, indicates another aristocratic connection.

But for Beethoven the chief excitements of this year may have been outside Bonn itself. As Grand Master of the Teutonic Order, the elector had to preside for many weeks over its sessions at Mergentheim, and he saw to it that he had his orchestra with him. The players' journey up the Rhine was accompanied by much revelry and clowning; in later years Beethoven retained many happy memories of this, as well as one curious memento (a mock diploma). An ambitious series of concerts was given at Mergentheim, and Beethoven

1. Autograph MS
of Beethoven's
scena and aria
'Primo amore'
WoO 92,
composed
c1790

also seized the opportunity of going with friends to Aschaffenburg, a summer palace of the Electors of Mainz, to visit the famous pianist Sterkel. It is said that Sterkel's light touch and graceful, fastidious style were a revelation to Beethoven. But when Sterkel challenged him to play his own Righini variations, doubting his ability to do so, it was Beethoven's turn to cause amazement, particularly since he improvised extra variations in a style that imitated Sterkel's.

By this time, it is clear, it was not only other professional musicians who recognized his worth or valued his friendship. He had formed a considerable circle of friends, drawn from some of the most discerning, progressive and respected families in Bonn. A few at least deserve mention here. Count Waldstein, eight years older than Beethoven, had come to Bonn from Vienna in 1788. A close associate of the elector and highly musical himself, he proved a devoted friend and patron of Beethoven, whom he came to know in the cultivated circle of the von Breuning family. Frau von Breuning, whose husband had died in a fire in 1777, had four children, all slightly younger than Beethoven: Eleonore, later to marry another friend of Beethoven's Bonn and early Vienna years, Franz Gerhard Wegeler; Christoph; Stephan, a lifelong friend; and Lorenz, who died young. The young widow herself became something of a second mother to Beethoven and seems to have had a keen insight into his character. She used her authority to dissuade him from neglecting duties that he found tedious, while evidently recognizing his tendency to self-absorption, since she would often remark: 'He has his raptus again'. She exercised some control, too, over his friendships; of the less suitable ones he remarked in later years: 'She understood how to keep insects off the flowers'. This kindly supervision, and the provision of what became almost a second home, meant much to Beethoven, who in spite of his many admirers remained in some ways a solitary youth, and on occasion a painfully shy one.

There were other opportunities for agreeable social life in Bonn. The elector was often absent, leaving Beethoven free for musical activities unconnected with the court. He spent much of his time in a circle of aristocratic friends and prosperous citizens such as the Westerholts, the Eichhoffs and the Kochs. The Kochs ran a kind of social and political club, the Zehrgarten, that was a centre for intellectual life in Bonn, and a number of Beethoven's early compositions were written for members of this circle.

It may have been Waldstein whose voice was decisive in the proposal that Beethoven should now go to Vienna to study with Haydn. When Haydn had passed through Bonn on his way to England in December 1790 he had met some of 'the most capable musicians', but it is not known whether Beethoven was among them. But in July 1792, according to Wegeler, the electoral orchestra assembled at Godesberg to give a breakfast for Haydn, now on his journey back to Vienna, and Wegeler adds that on this occasion Beethoven showed him a cantata (doubtless WoO 87 or 88) and received Haydn's commendation. More probably that had happened earlier, on Haydn's outward journey. But it was now that the matter of Beethoven becoming

Haydn's pupil was no doubt raised; the elector, to whom it fell to pay for the journey and the living expenses in Vienna, in due course sanctioned the arrangement. Beethoven's departure was fixed for the beginning of November. An *album amicorum* from this time records the good wishes of a large number of his friends, who had no reason to expect that he would be leaving Bonn for ever. None of the entries was more prophetic than that of Waldstein:

Dear Beethoven: You are going to Vienna in fulfilment of your long-frustrated wishes. The Genius of Mozart is still mourning and weeping over the death of her pupil. She found a refuge but no occupation with the inexhaustible Haydn; through him she wishes once more to form a union with another. With the help of assiduous labour you shall receive *Mozart's spirit from Haydn's hands*. Your true friend, Waldstein.

III 1792–5

Beethoven arrived in Vienna, the city that was to be his home for the rest of his life, in the second week of November 1792. He was not quite 22. His entry into Viennese circles was unobtrusive, and the sporadic entries in the little diary that he had started on his journey and kept at least until 1794 are the best guide to his immediate preoccupations. They show him looking for a piano and for a wig-maker, buying clothes, noting the address of a dancing-teacher, and the like. Later entries are concerned with the renting of some lodgings. And on the same page that records 'on Wednesday, 12 December [1792], I have 15 ducats', there is a variety of small sums of money set against the name of 'Haidn'. Within weeks of his arrival, therefore, the instruction from Haydn which had been the purpose of his journey had already begun. Of another event of the same month, the death of his father in Bonn on 18 December, there is no mention in the diary.

Haydn's tuition lasted for no longer than about a year; in January 1794 he left Vienna for his second London visit. The arrangement proved a disappointment to Beethoven, but he concealed this at the time from Haydn, and throughout 1793 the relations between pupil and teacher were outwardly cordial. Haydn appears to have had no corresponding misgivings – at any rate until later, when Beethoven had some very harsh things to say about him. Temperamentally, however, they were set for conflict. The childless Haydn no doubt wished for affection and even love from his most brilliant pupil – but that was the one thing that Beethoven was too mistrustful to give. Though he could write to the only moderately gifted (and no longer present) Neefe, 'If ever I become a great man, yours will be some of the credit', he was almost bound to feel the genius of 'Papa' Haydn standing in his way, one more father to be defied or circumvented. Beethoven's unease crystallized into the groundless suspicion that his teacher 'was not well minded towards him' and was neglecting or perhaps even sabotaging his tuition. (The formal side of the instruction can be seen from the surviving exercises, which consist of strict species counterpoint; they are in Beethoven's handwriting, with somewhat intermittent corrections by Haydn.) The lack of thoroughness on Haydn's part formed one of Beethoven's grievances. Finding that several of his errors had been overlooked, he enlisted the help for some months of another teacher, the composer Johann Schenk, but concealed this fact from Haydn.

It is not clear whether Haydn also instructed him in free composition. A clue here is provided by an episode that seems to reflect better on Haydn than on his pupil. Since leaving Bonn Beethoven had found himself with insufficient money for his living expenses. He continued, it is true, to receive his Bonn salary each quarter, and after his father's death he had successfully petitioned the elector to double it; but some part of this must have gone to support his brothers, who were still in Bonn. For his subsistence in Vienna he had only 100 ducats (nearly 500 florins) per annum. He had hoped to receive the whole of it on his arrival in Vienna, at which time he had to make considerable outlays, but it seems to have been paid quarterly. The result was that he had to borrow. On 23 November 1793 Haydn wrote on his behalf to the elector, enclosing five pieces of music, 'compositions of my dear pupil Beethoven', whom he predicted would 'in time fill the position of one of Europe's greatest composers'. He added (with characteristic generosity): 'I shall be proud to call myself his teacher; I only wish that he might remain with me a little while longer.' Haydn's letter next turned to the question of Beethoven's subsidy; it described the elector's 100 ducats as a sum quite inadequate to Beethoven's needs, pointed out that he himself had had to lend him 500 florins, and ended by suggesting that the elector might do well to increase the subsidy to 1000 florins in the coming year. The elector's reply was both accurate and icy. Four of the five submitted works had been composed and performed in Bonn long before the move to Vienna, and were therefore no evidence of progress. Moreover Beethoven was being paid not only the 100 ducats but also his ordinary salary of 400 florins, so had no reason to be in particular difficulty. The elector concluded:

I am wondering if he would not do better to begin his return journey, in order to resume his duties here; for I very much doubt whether he will have made any important progress in composition and taste during his present stay, and I fear he will only bring back debts from his journey, just as he did from his first trip to Vienna.

It looks as though Beethoven had misled Haydn in respect both of his total income and of the music that he had written in Vienna, and thus exposed Haydn to the elector's withering reply. This suggests in turn that Beethoven completed hardly anything new under Haydn's immediate supervision, though he seems to have revised and polished several of the later Bonn works.

A worsening relationship, however masked, may also explain why Beethoven did not accompany Haydn to England. He turned to another tutor, Johann Georg Albrechtsberger, the Kapellmeister at St Stephen's and the best-known teacher of counterpoint in Vienna. The lessons, three times a week, started after Haydn's departure and continued throughout 1794 to the spring of 1795. They were more thorough-going than Haydn's had been, and covered not only simple counterpoint but contrapuntal exercises in free writing, in imitation, in two-, three- and four-part fugue, choral fugue, double counterpoint at the different intervals, double fugue, triple counterpoint and canon – at which point they were broken off. Albrechtsberger proved a most conscientious, though at the same time very dry, teacher.

A third name is often linked with Haydn's and Albrechtsberger's: that of the imperial Kapellmeister Antonio Salieri. It was Salieri's genial custom to offer free tuition to impecunious musicians, especially in the setting of Italian words to music; and it is usually stated that Beethoven availed himself of this informal help soon after his arrival in Vienna. The only surviving evidence of any serious study with Salieri, however, dates from the years 1801–2, when he set a large number of unaccompanied partsongs with Italian words and a scena and aria for soprano and string orchestra (WoO 92*a*). These were followed in 1802 by two final pieces scored for orchestra, the terzetto *Tremate, empi, tremate* (op.116) and the duet *Ne' giorni tuoi felici* (WoO 93). They are more than exercises and may have been intended for a concert. In spite of Salieri's help Beethoven never fully mastered Italian prosody, though something had no doubt been gained in the skill of setting words by the time that he turned in the direction of opera.

But that is to jump far ahead. Aside from his studies, Beethoven's first task in Vienna was to establish himself as a pianist and composer. And that was something that he achieved both rapidly and with remarkable success. His gifts apart, there were at least two reasons for this, and they not only helped to launch him but continued to sustain him after he had gained an ascendant position. The first was his immediate contacts with aristocratic circles. He had arrived from

Bonn as the court organist and pianist to the Emperor Franz's uncle, and with a reputation already spread by high-born Viennese who had heard him while visiting the elector; he was a protégé of Count Waldstein, who was connected by birth or by marriage with several of the greatest houses of the Austrian, Bohemian and Hungarian nobility; and he was the pupil of Haydn. Thus he was in the strongest possible position to be introduced into the best aristocratic circles.

The second reason had to do with the character of the circles themselves. The aristocracy based on the Austrian capital surpassed all others of Europe in its devotion to music, and much of its time and a considerable part of its fortunes – a ruinous amount in some cases – was spent in the conspicuous indulgence of this taste. Not only did these aristocrats welcome virtuosos to their town palaces and country estates, but some of them, such as Prince Lobkowitz, kept private orchestras and even – like the Esterházys – opera companies as well. If their support was not on quite so lavish a scale, at least they employed a wind band or, like Prince Karl Lichnowsky and the Russian Count Razumovsky, a quartet of string players. The Court Councillor von Kees was among the many who organized private concerts; a large library of music was assembled by the Baron van Swieten, a patriarch whose distinction it was to cultivate the music of Bach and Handel and introduce it to Viennese audiences. The names of van Swieten and some of these others are found in the records of Mozart's and Haydn's lives; and they now gave a welcome to Beethoven.

He certainly needed more than their mere approval. His salary from Bonn was paid only till March 1794, and in a list of the elector's musicians from the autumn of the year he was entered as 'Beethoven, without salary in Vienna, until recalled'. (The elector now had his own difficulties as a result of the military victories of the neighbouring French. He had visited Vienna in January 1794, and Beethoven may have called on him and discussed his position.) Since many of the aristocracy had spacious accommodation or several houses, it was natural for them to provide Beethoven with lodging. One of the first houses in Vienna (if not the very first) in which he had rooms was owned by Prince Lichnowsky, who soon established himself as a leading patron of the composer. Both he and his wife Princess Christiane (née Thun) were intensely musical, and lavished a steady stream of kindnesses on him. But others were scarcely less generous or hospitable, so that it is no surprise to find Beethoven setting off in June 1793 for Eisenstadt, where Haydn was staying; doubtless the Esterházys looked after him. Another early supporter who became a lifelong friend was the Hungarian Nikolaus Zmeskall von Domanovecz. A capable amateur cellist and composer of quartets, he ardently promoted performances of Beethoven's music and continually rendered him small services, including the provision of quill pens, which Beethoven could never cut properly himself.

Beethoven's instant and striking successes as a virtuoso were at first confined to performances in private houses. Regular public concerts of the sort given throughout the season in London and Paris were

17

not then a feature of Viennese musical life; there were only a few annual charity concerts and an occasional subscription concert of a virtuoso or Kapellmeister. But in the salons the stunning effect of Beethoven's solo playing, and particularly perhaps of his improvising, was immediately recognized. A glimpse of what this aspect of his life was like to Beethoven is to be found in one of his letters to Eleonore von Breuning in Bonn, to whom – because of a quarrel before his departure from there – he did not write until he had been in Vienna for almost a year. He had dedicated to her the first of his works to be published in Vienna (composed in part in Bonn), his variations for violin and piano on Mozart's 'Se vuol ballare' (WoO 40), and in alluding to the difficult trills in the coda confessed to her:

I should never have written down this kind of piece, had I not already noticed fairly often how some people in Vienna after hearing me extemporize one evening would next day note down several peculiarities of my style and palm them off with pride as their own. Well, as I foresaw that their pieces would soon be published, I resolved to forestall those people. But there was another reason, too; my desire to embarrass those Viennese pianists, some of whom are my sworn enemies. I wanted to revenge myself on them in this way, because I knew beforehand that my variations would here and there be put before the said gentlemen and that they would cut a sorry figure with them.

The pugnaciousness of the virtuoso is characteristic, and it was not long before he displayed his powers before wider audiences.

An early opportunity came at a charity concert in the Burgtheater on 29 March 1795. Beethoven appeared as composer as well as virtuoso, for he played a piano concerto of his own, probably the work in B♭, later published as the Second Concerto (op.19). His old friend from Bonn, Franz Gerhard Wegeler, who was in Vienna from October 1794 to the summer of 1796, witnessed the preparations for this concert – or it may have been the one nine months later in December and the concerto may have been the First (op.15) in C – and relates how Beethoven completed the finale only at the very last moment while suffering from severe abdominal pains. At a second charity concert the next day Beethoven again appeared on the platform; this time he gave an improvisation. And on 31 March he played for the third time in three days at a performance of Mozart's *La clemenza di Tito* organized by his widow; this time the concerto was one of Mozart's.

Apart from the variations dedicated to Eleonore von Breuning he had not yet published anything in Vienna. The decision was deliberate, for his op.1 was intended to be an event. He chose a set of three piano trios, a genre dear to aristocratic devotees of chamber music, and he dedicated it to Prince Lichnowsky. The trios had already been heard and admired, possibly in earlier versions. There is a well-known story of what purports to have been their first performance at a soirée of Lichnowsky's at which Haydn was present; although he praised them, he is said to have advised Beethoven not to publish the third of them, in C minor. If this story is true down to the details, the soirée must have taken place before Haydn's departure for England in January 1794, for when he returned to Vienna in August 1795 op.1 had just been published. But it seems more likely that he heard the trios only on his return, and expressed regret about the inclusion of the C minor one. Since the third trio ultimately proved the most successful, Beethoven suspected malice on Haydn's part; years later Haydn confirmed that he had had misgivings about its publication, adding that he had not believed it would be understood and received so well. Beethoven published his op.1 by subscription, the edition being produced by the publisher Artaria. The subscription list contained 123 names, and the subscriptions amounted to 241 copies at one ducat (roughly four and a half florins) each; since Beethoven paid the publisher only a florin per copy he made a handsome profit.

Haydn's return to Vienna was marked by the performance at Lichnowsky's of another substantial composition by Beethoven: the three piano sonatas that he subsequently published in March 1796 as his op.2 and dedicated to Haydn. It is said that Haydn had hoped Beethoven would append to his name on the title-page the words 'pupil of Haydn' – a common enough custom – and that Beethoven declined to do so, privately declaring that although he had had some lessons from Haydn he had never learnt anything from him. At all events the sonatas were published in March 1796 without any acknowledgment of pupillage.

Outwardly, however, relations between the two did not appear to be badly strained. On 18 December 1795 Beethoven made his second public appearance in Vienna as a composer-virtuoso, playing a piano concerto at a concert which Haydn organized and which included three of his latest symphonies, written for London. It is probable that this was the first performance of the C major concerto. Another sign of

Beethoven's growing popularity was the invitation this year to write the minuets and German dances for the November ball held in the Redoutensaal by the Pensionsgesellschaft Bildender Künstler.

IV 1796–1800

Beethoven's sights were now set on still wider audiences. His youngest brother Nikolaus Johann had arrived from Bonn at the very end of 1795 and had found employment in an apothecary's shop; and Caspar Carl, the other brother, had been in Vienna from the middle of 1794, apparently supporting himself by giving music lessons. Thus all three brothers were united in Vienna, and Beethoven now felt able to embark on a concert tour. In February 1796 he set out for Prague, travelling (as Mozart had done seven years earlier) with Prince Lichnowsky. Writing from Prague to his brother Johann in Vienna he announced his intentions of visiting Dresden, Leipzig and Berlin, and added: 'I am well, very well. My art is winning me friends and respect, and what more do I want? And this time I shall make a good deal of money'. On 11 March he gave a concert in Prague; on 29 April he played before the Elector of Saxony in Dresden. On reaching Berlin, he appeared several times before the King of Prussia (Friedrich Wilhelm II), and with the king's first cellist, Jean Louis Duport, he played the two op.5 cello sonatas, written for this performance. Another *pièce d'occasion* was the set of 12 variations for cello and piano on a theme of Handel; the cello was of course the king's instrument, and the choice of theme ('See the conqu'ring hero comes') may have contained a courteous nod towards the throne. The king gave Beethoven a gold snuffbox filled with louis d'ors: 'no ordinary snuffbox', Beethoven later declared with pride, 'but such a one as it might have been customary to give to an ambassador'. He seems to have stayed for about a month in Berlin, making the acquaintance of the Kapellmeister, Himmel, as well as of Zelter and Fasch, and twice giving improvisations before the Singakademie.

By the time that Beethoven returned to Vienna his friend Wegeler had gone back to Bonn, together with Christoph von Breuning, though Christoph's brother Lorenz remained in Vienna. Beethoven and Wegeler – who completed his studies in medicine, married Eleonore von Breuning in 1802, and set up practice in Koblenz – never met again, but they remained friends and exchanged letters from time to time. Wegeler's contribution to the *Biographische Notizen über Ludwig van Beethoven* that he compiled with Ferdinand Ries after Beethoven's death and published in 1838 (with a supplement, 1845) is a valuable source of information on Beethoven's childhood and adolescence in Bonn and on his life in Vienna up to 1796.

At the end of 1796 Beethoven again travelled. He played at a concert at Pressburg (now Bratislava) on 23 November. The next year, 1797, is almost devoid of incidents that have left any record. At the end of May he wrote to Wegeler that he was doing well – in fact, better and better; on 1 October he penned some warm lines in the album of Lorenz von Breuning, who was leaving Vienna to return to Bonn. Between those dates nothing is known, and it is even possible that he was seriously ill at that time. One source assigns such an illness to the second half of the previous year, where there is also a gap in the records (from July to November). The year 1797 saw the publication of several compositions: his opp.5–8, the most important of which were the E♭ Piano Sonata (op.7) and the cello sonatas written for Berlin (op.5), as well as the song *Adelaide* (op.46), dedicated to the author of its words, the poet Matthisson. The publications of 1798 were even more assured, including the three op.9 string trios, his most impressive chamber works to date, and the three op.10 piano sonatas. The trios were dedicated to Count Johann Georg von Browne, a patron whom Beethoven described in the dedication as the 'first Maecenas of his Muse', while op.10 was dedicated to Browne's wife.

Early in 1798 considerable interest was aroused by the arrival in Vienna of the emissary of the French Directoire, General Bernadotte; in his retinue was the violinist Rodolphe Kreutzer. Both were only a few years older than Beethoven, whose acquaintance they made. Bernadotte's sojourn in Vienna was brief, but he is said to have suggested to Beethoven the idea of writing a 'heroic' symphony on the theme of the young General Bonaparte.

Later in the year (the exact date is unknown) Beethoven visited Prague and gave two public concerts, as well as a private recital. They were attended and described in some detail by the Bohemian composer Václav Tomášek (Wenzel Tomaschek). He heard Beethoven play the Adagio and Rondo from the Piano Sonata in A op.2 no.2, improvisations on 'Ah perdona' from Mozart's *Tito* and on 'Ah vous dirai-je maman', and both the B♭ and C major piano concertos (Tomášek described the former as having just been written for Prague, so it was probably a revised version that was performed). For Tomášek, who by the end of his life had heard all the outstanding virtuosos from the age of Mozart to the 1840s, Beethoven remained the greatest pianist of all – though Beethoven the composer came in for more criticism. Only in 1798–9, in fact, did Beethoven's virtuosity, which seems until then to have had no serious rivals in Vienna, come under challenge from the Salzburg-born pianist Joseph Wölfl and from Johann Baptist Cramer of London; both were about his age. The stimulus of competition from two such excellent players, whose strengths were nevertheless rather different from his own, could only have had a salutary effect on his playing, which he was to describe in 1801 (to a correspondent who had not heard him for two years) as having 'considerably improved'.

It was probably a living composer whose challenge Beethoven was finding more dispiriting. In 1795–6 he had reacted to the brilliant symphonies that Haydn had brought back from London by attempting to write a symphony of his own in C major, but although he worked at it vigorously it remained unfinished and was abandoned. Now, in April 1798, Haydn gave a private performance of his new oratorio *Die Schöpfung (The Creation)*, and Beethoven might well be excused for believing his old teacher's confession that the inspiration for some passages was more than human. Furthermore, Haydn continued to produce masterly string quartets with unabated vigour: six had been written in 1793 and six more in 1797. Although all the works with opus numbers that Beethoven had so far published in Vienna, apart from the piano sonatas, could loosely be called chamber works, the particular genre that was most closely associated with Haydn, and indeed with Mozart as well – the string quartet – was noticeably unrepresented. That Beethoven was only too aware of their formidable example there can be no doubt, and he copied out movements from several of their quartets in score for closer study. Still, the challenge was one for which he now felt himself ready, and in the second half of 1798 and through the winter and spring he worked on a set of quartets.

It is tempting to draw a connection between the self-consciousness of this undertaking and a change in his working methods which coincided with it. Beethoven had always made sketches of the compositions that he was engaged in writing, and as time went on they became more voluminous. But hitherto they had been written on loose single leaves or bifolia of music paper. From the middle of 1798 he began to make his sketches in books of music paper. The first two of the sketchbooks contain sketches for four of the quartets that he was now writing, as well as for a considerable number of other works that he completed, revised or attempted to write in the same months. (The completed works include a song, *La tiranna* WoO 125, which he wrote to English words, working in part from a phonetic transcription.) The sketchbooks evidently retained some value for him long after they had been filled up, for he kept them by him and preserved most of them in a growing pile for the rest of his life. Some aspects of

25

their importance, a particular preoccupation of Beethoven scholarship in recent years, are discussed below on page 154.

In 1798 Karl Amenda, a student of theology and a competent violinist, arrived in Vienna from his native Courland (Latvia), and became tutor to Prince Lobkowitz's children and music teacher at the home of Mozart's widow. He and Beethoven soon became fast friends; indeed they were almost inseparable. But in the late summer of 1799 Amenda was obliged to depart again for Courland, and on 25 June 1799 Beethoven gave him a copy of a quartet 'as a small memorial of our friendship'. This quartet was later published in a somewhat altered form as the first of the op.18 quartets. It is not clear how many of the six quartets had been completed by the end of 1799; but the ones written first were in any case revised later before being sent to the publisher.

Other friendships formed around this time were ultimately more fateful for Beethoven. In May 1799 the Countesses Therese and Josephine von Brunsvik, then 24 and 20, came to Vienna from Hungary on a short visit with their widowed mother, who wished them to take lessons from Beethoven. He was charmed by them, proved a very attentive teacher, and for their album composed a 'musical offering' consisting of a song with some variations for piano duet (WoO 74). Through them he became friends with the other members of the family, their brother Franz and their youngest sister Charlotte; Julie (Giulietta) Guicciardi, who came to Vienna from Trieste with her parents in 1800, was their very young cousin. Beethoven was soon a welcome guest on visits to their estates in Hungary. But the short trip to Vienna had unhappy consequences for Josephine. The family made the acquaintance of Count Joseph Deym (or Herr Müller; he had been exiled after a duel and returned under a pseudonym); Deym was the proprietor of a famous museum of waxworks, and although he was almost 30 years older than Josephine, her mother pressed his claim as a suitable husband for her, partly no doubt in an attempt to redeem the family fortunes. Josephine reluctantly assented, and they were quickly married; but Deym was in fact badly in debt, so that even financially the match had nothing to be said for it. The visits of Beethoven to the wing of the 80-room museum house in Vienna in which the Deyms lived must have afforded some consolation to the unhappy young countess.

On 2 April 1800 Beethoven gave his first concert for his own benefit, in the Burgtheater. The music included, besides a Mozart symphony and numbers from Haydn's *Creation*, two new works by Beethoven, the Septet (op.20) and the First Symphony. The former soon became one of his most popular works; the reception of the latter was appreciative, although the heavy scoring for the wind was remarked on. His piano playing was on display in an improvisation and a piano concerto – either the C major or the B♭. No doubt he had planned to produce a new concerto, the Third in C minor, written around this time but not performed until the spring of 1803 (the score, with a heavily revised solo part, is dated '1803'). Perhaps, then, the C minor concerto could not be completed in time for the concert. For his appearance later in the same month with the Bohemian horn player Johann Wenzel Stich (or 'Giovanni Punto', the name that Stich preferred to use) he very rapidly wrote a horn sonata (op.17); they gave a second concert three weeks later in Pest. Beethoven may have spent part of the summer of 1800 with the Brunsvik family in Hungary.

The second half of 1800 was outwardly uneventful; it doubtless saw the final revision of the op.18 string quartets, and the writing of the B♭ Piano Sonata (op.22) and of the A minor and F major violin sonatas (opp.23, 24). There was less inducement to prepare new works for a possible concert in the following spring, since he had received an important commission for the court stage: he was to write the music for a ballet designed by the celebrated ballet-master Salvatore Viganò, *Die Geschöpfe des Prometheus* (op.43). This was given its first performance at the Burgtheater on 28 March 1801 and was successful enough to be repeated more than 20 times. Only a sketch of the scenario survives. In the finale Beethoven used a melody that evidently came to assume a certain emotional importance for him, perhaps even embodying something of his spirit of determination and heroism in battling against difficulties, for he used it again as the theme for two important and challenging sets of variations completed in 1802 and 1803: the op.35 piano variations and the variation-finale of the Eroica Symphony.

By this time several publishers were competing for Beethoven's newest works, and though a number of important compositions had lately appeared – the highly individual *Sonate pathétique* (op.13), for in-

28

stance, dedicated to Prince Lichnowsky, at the very end of 1799 – others had not yet found a buyer. An entertaining correspondence with the publisher Franz Anton Hoffmeister, who had lately moved from Vienna to Leipzig, dates from around this time. Hoffmeister finally bought several works beginning with the First Symphony, the Second Piano Concerto, the Septet and the B♭ Piano Sonata.

It comes as a surprise to find that Beethoven was intending to dedicate the symphony to his former overlord and employer, the Elector of Cologne. The preceding years had been harsh to Maximilian Franz. After being forced by French military successes to leave Bonn in October 1794, and having stayed for a while in various cities, he had finally returned to Vienna in April 1800 and settled in Hetzendorf just outside the city. Beethoven is believed to have spent some time in summer 1801 in and around Hetzendorf, and may well have called on the elector and paid his homage or made his peace with him, for the instructions for the symphony's dedication are contained in a letter to Hoffmeister written about 21 June 1801. Beethoven's wishes were not to be carried out, for the elector died on 26 July and the symphony was subsequently dedicated to Baron van Swieten.

V 1801–2; deafness

At a time of personal crisis it was natural for Beethoven's thoughts to turn to his last years in Bonn and to the friends he still had there. One of these – his friend of longest standing, trained in medicine, discreet, remote from Vienna – was particularly suited to be the

29

first recipient of a secret that Beethoven had kept to himself for some years and that had not yet been guessed by his circle of friends in the capital: the appalling discovery that he was going deaf. These tidings were now conveyed to Wegeler in Bonn in a letter of 29 June 1801, and to another absent friend, Karl Amenda in Courland, two days later.

Exactly when Beethoven first detected some impairment in his hearing cannot be determined. Perhaps he did not quite know himself, for no doubt its onset was insidious, and he probably did not regard any temporary periods of deafness or diminished hearing as sinister, especially since he had long become used to spells of fever, abdominal pain and episodes of ill-health. A young man does not expect to go deaf, and although in one account he implied that he had noticed the first symptoms in 1796, other statements set the date somewhat later, and the crisis came only with the growing realization that his deafness was progressive and probably incurable. From the descriptions of his symptoms there is general agreement among modern otologists that his deafness was caused by otosclerosis of the 'mixed' type, that is, the degeneration of the auditory nerve as well – by no means a rare condition.

At this time Beethoven had not yet given up hope that his doctors could do something for his hearing, but he could already foresee incalculable troubles both for his professional life and – what it is easy to forget was equally important to him – for his social life. As he wrote to Wegeler:

I must confess that I am living a miserable life. For almost two years I have ceased to attend any social functions, just because I find it impossible to say to people: I am deaf. If I had any other profession it would be easier, but in my profession it is a terrible handicap. As for my enemies, of whom I have a fair number, what would they say?

To Amenda he wrote in similar terms: 'Your Beethoven is leading a very unhappy life, and is at variance with Nature and his Creator', but he added that when he was playing and composing his affliction still hampered him least – it affected him most when he was in company. A curious feature of these letters, in fact, is that each includes not only a melancholy account of the despair which his deafness had brought about but also an almost lyrical portrait of his professional and financial successes. Lichnowsky had agreed to pay him an annuity of 600 florins for some years; six or seven publishers were competing for each new work; he was often producing three or four works at the same time; his piano playing had considerably improved: 'why, at the moment I feel equal to anything'.

Four and a half months later Beethoven again wrote at length to Wegeler: his doctors had been unable to help his hearing, but he was leading a slightly more pleasant life.

You can scarcely believe what an empty, sad life I have had for the last two years. My poor hearing haunted me everywhere like a ghost; and I avoided all human society. I was forced to seem a misanthrope, and yet I am far from being one. This change has been brought about by a dear charming girl who loves me and whom I love ... and for the first time I feel that marriage might bring me happiness. Unfortunately she is not of my class.

This letter is similar to the earlier ones in containing phrases that are very exalted in tone: 'I will seize Fate by the throat; it shall certainly not crush me completely – Oh it would be so lovely to live a thousand lives'. Such passages, and their more gloomy counterparts, are characteristic of his conflicting moods as he faced the prospect of permanent deafness and the quite unexpected threat to what had hitherto been a triumphant career. An attitude of pious resignation, with which he tried to master such unruly feelings, did not come easily to him but found expression in the six hymn-like settings of sacred poems by Gellert (op.48), which he completed at about this time.

The 'dear charming girl' who was brightening Beethoven's days was no doubt the Countess Giulietta Guicciardi. She was now not quite 17: too young, and perhaps too spoilt, to take Beethoven's devotion very seriously, though no doubt she was flattered for a time by the attentions of a famous composer, a man much admired by her cousins. Much, probably too much, has been made of the fact that it was to her that he dedicated the 'Moonlight' Sonata (op.27 no.2), written in 1801. But it is clear that for a time he was under her spell – she even boasted of this – and he must have had mixed feelings when in November 1803 she married Count Wenzel Robert Gallenberg, a prolific composer of ballet music, who was only a year older than herself.

By the end of 1801 Ferdinand Ries, the son of Franz Anton Ries who had befriended the Beethoven family in Bonn, was living in Vienna, and Beethoven agreed to take him as his piano pupil. Ries was then just 17 and he remained with Beethoven until the autumn of 1805, when he had to return to Bonn for military service.

During those four years he had unrivalled opportunities for observing Beethoven at his work, on his walks in the countryside, with his brothers and his friends, or at the social functions of the aristocracy. His recollections of this time, set down somewhat artlessly in the *Biographische Notizen* which he compiled in collaboration with Wegeler in the 1830s, are invaluable as an authentic and unsentimental picture of Beethoven. A recurring theme in Ries's account is Beethoven's unwillingness, or inability, to conform to the normal conventions of social punctilio, and especially to play the courtier and to oblige by performing to a private audience when requested to do so. These last attitudes, indeed, hardened in later life into a stance in which he felt himself a prince of art and entitled to behave as one.

One particular aspect of Beethoven's behaviour that obviously baffled Ries was his relations with his brothers: he was appalled to see grown men come to blows in the street in the middle of an argument. In ascribing to the scheming of his brothers many of the difficulties that Beethoven was experiencing both in his relations with friends and in his practical arrangements, Ries may have been loyally taking his teacher's side. There is no doubt, however, that Caspar Carl in particular then played an important part in Beethoven's business affairs. For several years, starting in 1802, he was entrusted with the offer of new compositions to publishers, and with the subsequent negotiations. But on 25 May 1806 he married Johanna Reiss, the daughter of a well-to-do upholsterer; their only child, Karl, was born on 4

September. After that Beethoven largely dispensed with his brother's help, but his nephew later assumed a position of great importance in his life.

The summer of 1802 was spent just outside Vienna in the village of Heiligenstadt. It was there, no doubt, that Beethoven put the finishing touches to the Second Symphony and completed several other works of this prolific year: the three op.30 violin sonatas, the op.33 bagatelles, and perhaps the first two of the op.31 piano sonatas. He had gone to Heiligenstadt in the spring, perhaps with the thought of spending longer in the country than usual for the sake of his health and hearing. Now in October, as he prepared to return to the city, he carefully wrote out a strange document addressed to his two brothers (though wherever his brother Johann's name was implied there was a blank space). Found among his papers after his death and known as the 'Heiligenstadt Testament', it is dated 6 October 1802 at the beginning and 10 October at the end, and its contents mark it as representing a trough of despondency in his fluctuating moods. His hearing had shown no improvement in the country, and he re-cognized that his infirmity might be permanent; he defended himself against the charge of misanthropy, and taking leave of his brothers declared that though he had now rejected the notion of suicide, he was ready for death whenever it might come. The Testament has always been recognized as a poignant witness to the despair that often overwhelmed Beethoven at this time.

VI 1803–8

From that nadir of despondency Beethoven seems to

have recovered quickly, and probably by his usual means: hard work. His next activities certainly indicate a firm repudiation of the notion that his deafness would handicap him professionally. Caspar Carl wrote to Breitkopf & Härtel on 12 February 1803:

You will have heard by now that my brother has been engaged by the Wiedener Theater [i.e. Theater an der Wien], he is writing an opera, is in charge of the orchestra, can conduct if necessary, seeing that there is a director already available there every day. He has assumed the chief direction mostly so as to have a chorus for his music.

Although Beethoven had already gained a reputation throughout Europe as a composer of instrumental music, opera was still the royal road to fame. At the time there was something of a dearth of local talent in opera at Vienna, but from the spring of 1802 the importation of operas from Paris had more than compensated for this. Those of Cherubini and to a lesser extent of Méhul became extremely popular; so great indeed was the clamour for Cherubini's music that one of his operas (*Les deux journées*) was staged at rival theatres on successive nights. Like the other Viennese, Beethoven responded enthusiastically to these operas from revolutionary France, with their contemporary realism and heroic plots (copied in some cases from recent political history). Thus he eagerly took up the invitation to write an opera for Schikaneder's theatre and moved his lodgings to the Theater an der Wien.

An immediate bonus for this appointment was the opportunity to give a concert. He quickly wrote his oratorio *Christus am Oelberge*, and it was performed

2. Ludwig van Beethoven: miniature (1803) by Christian Horneman

on 5 April 1803 together with the First and Second
Symphonies and the Third Piano Concerto (with
Beethoven as soloist), all but the First Symphony being
new to the audience. The oratorio, which tells of the
Agony in the Garden (and is known in English-
speaking countries as *The Mount of Olives*), marked
Beethoven's first appearance in Vienna as a dramatic
vocal composer. Another rapidly written piece was

occasioned by the arrival in Vienna of the young mulatto violinist George Polgreen Bridgetower: the Kreutzer Violin Sonata (op.47) was played by Bridgetower and the composer on 24 May. He may also have started to look at the opera, *Vestas Feuer*, with a libretto by Schikaneder.

But something else was evidently pressing: the inner demand to complete a great instrumental work. The writing of the Third Symphony, the *Sinfonia eroica*, was the major effort of the summer of 1803, which was spent in Oberdöbling. The symphony was originally entitled simply 'Bonaparte', in tribute to the young hero of revolutionary France, who was almost exactly Beethoven's age. But this idealization of Napoleon as a heroic leader gave way to disillusionment when the First Consul proclaimed himself Emperor in May 1804. The story of Beethoven's rage when the news of this reached Vienna is well known: he went to the table where the completed score lay, took hold of the title-page and tore it in two. On its publication in 1806 the symphony was given its present title of 'heroic symphony', and was described as having been 'composed to celebrate the memory of a great man'. It was not the only work by Beethoven from these years that appears to reflect or embody extra-musical ideas of heroism. A similar spirit pervades the so-called 'Waldstein' Sonata (op.53), for instance, composed immediately after the symphony in the last months of 1803, and the 'Appassionata' Sonata (op.57), begun in the following year. Even the string quartets of this period, the three of op.59 completed in the summer of 1806 and dedicated to Count Razumovsky, are cast in the same mould.

In comparison with the exhilarating work on these instrumental pieces, the opera dragged: by the end of 1803 Beethoven had completed less than two scenes of *Vestas Feuer*, and he abandoned it. For a more attractive operatic libretto had come his way, and was to capture his imagination to a profound extent. This was J. N. Bouilly's *Léonore ou L'amour conjugal*. The plot – the tale of a political prisoner's rescue from a Spanish Bastille, engineered by his wife disguised as a man – is said to have been based on a real incident in the French Revolution. At first, no doubt, Beethoven was drawn by the opportunity that it afforded of writing a grand 'rescue' opera similar to those of the admired Cherubini, and on 4 January 1804 he informed the Leipzig critic Rochlitz that he was beginning to work on it. But the profounder implications that the story held for his own psychology will have emerged as the labour progressed; oppressed and isolated by his undeserved deafness, it was easy for him to identify with the unjustly imprisoned Florestan who lay alone in the dark with no apparent hope of rescue. (In the same way Christ's 'cup of sorrow' in the oratorio of 1803 seems to have been linked in Beethoven's mind with his own affliction.) And it was surely another side of his nature that could feel empathy with the spirited and ever-devoted Leonore; sustained by her vision of hope and longing, and following her 'inner drive', she is in some ways an even more Beethovenian figure than Florestan.

A change in the ownership of the Theater an der Wien in February 1804 rendered void Beethoven's contract to write an opera for the house. It may also have obliged him in due course to find new lodgings; at all events he arranged to share rooms with Stephan von Breuning, but a serious quarrel – induced mainly by Beethoven, it seems – broke out between the two friends, and by July Beethoven had moved for some weeks to Baden, a resort some 16 miles south of Vienna. Breuning reacted philosophically and with forbearance, and in November wrote to Wegeler, who knew them both well: 'You cannot conceive what an indescribable, I might say fearful, effect the gradual loss of his hearing has had on him'. The breach was made up, but in spite of some strenuous events, such as the first (private) performance of the Eroica, this was not a happy summer for Beethoven, and for a time he may have thought of leaving Vienna altogether – perhaps for Paris. But towards the end of the year his contract for the opera was renewed, and he set to work on it again.

Apart from the opera there was another reason for Beethoven to remain in Vienna. In January 1804 Count Deym, the husband of Josephine von Brunsvik, had died; the young widow, who now had four small children, continued to spend much of her time in Vienna, and by the autumn Beethoven, who had remained in touch with the family, became a frequent visitor to the house. He gave Josephine piano lessons. An intense relationship soon developed between them, the nature and course of which must be inferred from the contents of 13 letters that Beethoven wrote to Josephine between the autumn of 1804 and the autumn of 1807, and from drafts of some of her replies (these documents were first published in 1957). Beethoven, it

39

is clear, was passionately in love; Josephine, though moved by his devotion and keenly concerned with his happiness, his ideals and his art, retained a certain reserve throughout and rejected any intimacy closer than that of warm friendship. It would not be hard to find reasons why, after one unhappy marriage and with a young family now claiming her concern, she should be reluctant to throw in her lot with someone of Beethoven's uncontrolled nature, his want of much that passed for conventional good manners, and his unimpressive social standing. In the view of her sentimental unmarried sister Therese (writing many years after these events) it was consideration for her children that proved the decisive factor with Josephine. But a social barrier surely worked to keep the pair apart as well; it is noteworthy both that they were anxious to conceal the extent of their intimacy from the Brunsvik relatives, and that in addressing each other they used the formal 'Sie', not the more intimate 'du', which he kept for her brother, Count Franz Brunsvik.

The most intense period of the relationship was at the end of 1804 and in the first months of 1805: close to the time at which Beethoven was composing the triumphant finale of his opera, a paean to the accomplishments of a virtuous wife and to 'married love'. It came to an end by the autumn of 1807, with rueful scenes and misunderstandings, and with Beethoven still asking for closer contact than Josephine was prepared to concede. The following summer she left Vienna, and in 1810 married a Baron von Stackelberg; her second marriage, like her first, was not a happy one. She died in 1821.

By the summer of 1805 the opera was complete, but censorship difficulties postponed its first performance until 20 November. This had unfortunate consequences for its success, for in the preceding weeks the conquering French armies were advancing on Vienna. On 9 November the empress departed, and four days later Napoleon's troops entered the city. Thus the audience for the opera's first night consisted not of the Austrian nobility and moneyed classes, Beethoven's natural supporters and admirers, who had mostly fled from the capital, but of a miscellaneous crowd that included a sprinkling of French officers. Its reception was not enthusiastic, and after the third performance it was dropped. But with the return of Beethoven's friends to Vienna and the resumption of normal conditions there was pressure for the opera's revival, though also a general agreement that it had failed in part from its excessive length and in particular from the slowness of some of the earlier scenes. Beethoven was persuaded to make drastic cuts, which he did only with the greatest reluctance, for while some of these undoubtedly speeded the dramatic pace, others were mutilating. For the new version he provided an overture, *Leonore* no.3, which was itself a revision of the first production's overture (*Leonore* no.2). In its altered form the opera was now given two performances (29 March and 10 April 1806); then Beethoven was involved in a dispute with the director of the theatre, Baron Braun, and withdrew his score. It was not for another eight years that the opera was again seen on the stage. It had always been Beethoven's intention for the opera to be known as *Leonore*; but in both 1805 and

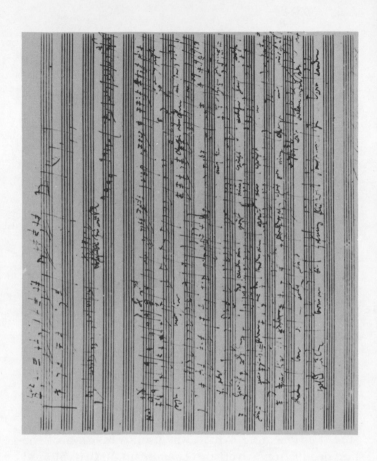

3. Autograph sketches for the Prisoners' Chorus from Act 2 of 'Leonore' (1805 version)

1806 it was billed as *Fidelio*, and for the 1814 production (see below) he acquiesced in that name. (The title *Leonore* is nowadays often used to distinguish the 1805 and 1806 versions from the more familiar 1814 one.)

The twin distractions of his opera and of his love for Josephine, and perhaps (at a deeper level) his slow adjustment to the fact of his deafness, may have led to some falling off in the quantity of new compositions during 1804 and 1805. But the period from the spring of 1806 to the end of 1808 must be regarded as one of prodigious fertility, with a steady stream of completed works, many of them on the largest scale. A comment that he wrote down among the sketches that date from the summer of 1806, some of which was spent in Silesia at the country seats of Prince Lichnowsky and Count Oppersdorff, reveals something of his optimistic and resolute mood: 'Just as you plunge yourself here into the whirlpool of society, so in spite of all social obstacles it is possible for you to write operas. Your deafness shall be a secret no more, even where art is involved!'.

Among the works completed before the end of the year were the three string quartets dedicated to the Russian ambassador Count Razumovsky, the 'Appassionata' Sonata, some at least of which had been composed earlier, the Fourth Symphony, the Violin Concerto, and in all essentials the Fourth Piano Concerto. They were quickly introduced to the public. The Violin Concerto, a work completed very rapidly, was performed by Franz Clement on 23 December 1806, and the Fourth Symphony and Fourth Piano Concerto were included at two concerts given at the palace of Prince Lobkowitz in March 1807, together with a new overture, to Collin's tragedy *Coriolan*. A further overture, apparently written for a planned production of his opera at Prague, was also composed around this time but never performed in public; it came to light only after Beethoven's death and is now known by the illogical title of '*Leonore* no.1' (op.138).

With the exception of the first two Razumovsky quartets, which were at first found 'difficult', these great works delighted the discerning Viennese audiences and enhanced Beethoven's fame throughout Europe. There were many signs of this. In April 1807 Muzio Clementi, now head of a prominent London firm of music publishers and piano makers, called on Beethoven in Vienna and secured the exclusive English rights to some of his newest compositions. And, nearer home, he received an invitation from Prince Nikolaus Esterházy II, Haydn's last patron, to produce a mass in celebration of his wife's name day in September 1807. This was a commission that made Beethoven unusually nervous. The type of composition required was not merely one in which he was inexperienced; it was one that had been mastered with special excellence by Haydn, who in the years up to 1802 had written six such masses for the princess's name day. Comparisons between Haydn's works and that of his one-time pupil were therefore inevitable. And in the event the Mass in C (op.86) was not well received, though Beethoven himself regarded it highly. After passing the summer in Baden working on the mass, he went to Eisenstadt for its first performance, on 13 September; later he spent some time at Heiligenstadt, no doubt completing the Fifth Symphony and his A major Cello Sonata (op.69) in the next few months. Some of the ideas for the symphony had been jotted down as early as the first months of 1804, but 1807 was the year that the main writing was done – and probably not before the mass was out of the way. Nor was there any slackening in the pace of composition in the next year, 1808. In fact that summer (which he again spent at Heiligenstadt) saw the writing of one of his largest and most characteristic works, the Sixth Symphony, called *Sinfonia pastorale*. He followed this directly with the two op.70 piano trios.

Yet behind all this flurry of creative activity there was one problem to which Beethoven had not yet found a satisfactory solution. He had no regular or dependable source of income. He could of course count on the generosity of the aristocratic circles that continued to admire him, on the fees payable for dedications, and on the sales of his music to publishers. Yet this was little enough to rely on; he was, after all, living in the city in which Mozart had died in poverty a decade and a half earlier, partly no doubt from having no adequately paid position. It was not easy for him to arrange a concert from which he could secure the receipts, since most concerts were private aristocratic affairs, or they were given for charity – at which Beethoven usually offered his services. There was occasionally the opportunity of obtaining one of the theatres for a benefit performance at a time when they were otherwise closed (Holy Week or around Christmas), but this often led to disappointments – in

1802, for instance, and again in 1807. In the latter year, therefore, he petitioned the Directors of the Imperial Theatres for a commission to compose an opera every year, for an income of 2400 florins; and he urged strongly his claim, whether this petition was granted or not, for an annual benefit day at one of the theatres. The petition contained a hint that otherwise he might have to leave Vienna. The reply (if any) of the Directors has not survived. No operatic commission followed, but after several postponements the Theater an der Wien was finally put at his disposal for the night of 22 December 1808, partly in just recognition of his services to charity; so he arranged to give an enormous benefit concert.

The working out of that evening contained many features characteristic of Beethoven. The programme was injudicious, consisting as it did of four hours of music, virtually all of it unfamiliar: first performances of the Fifth and Sixth Symphonies, and first public performances in Vienna of the Fourth Piano Concerto (with Beethoven as soloist) and portions of the Mass in C, as well as a piece written for Prague 12 years before, the scena and aria *Ah! perfido* (op.65); in addition Beethoven was to improvise. As if that were not enough for the audience, he decided the evening needed a finale; and since a chorus was already available, he rapidly threw together the work now known as the 'Choral Fantasy' (op.80). This consisted of an introduction for piano solo (extemporized by Beethoven at the first performance), several variations for piano and orchestra on a simple song melody that he had written in the 1790s, and a short choral conclusion.

Written at the last minute, the work was under-rehearsed; the orchestra, already on bad terms with Beethoven after a dispute in rehearsals for an earlier charity concert, broke down in the middle of the Fantasy and had to be restarted; Beethoven had quarrelled with the original soprano for the aria and her very young replacement was inadequate; and the theatre was bitterly cold. Thus the success of the evening was very mixed. The financial results are not known.

VII 1809–12

Even before the concert took place Beethoven had received the offer of a regular position: that of Kapellmeister at Kassel, where Napoleon's youngest brother Jerome Bonaparte, a youth in his early 20s, had been installed as 'King of Westphalia'. But although Beethoven usually had some sharp words for the Viennese, and continued to criticize them for the rest of his life, it is plain that he had no intentions of leaving Vienna if that could possibly be avoided. The Kassel appointment, with few obligations attached, was worth 600 ducats, plus 150 ducats travelling expenses: a total corresponding to about 3400 florins annually; moreover it was for life, or at any rate for as long as the 'king' retained his throne. Beethoven now used it to obtain a matching offer from Vienna. Although his initial conditions for remaining there included the guarantee of an annual concert and contained a strong desire for the title of Imperial Kapellmeister, their essence was a yearly salary of 4000 florins. And this after a month or two of negotiation he was able to obtain. A document dated 1 March 1809 guaranteed that its three signatories would provide Beethoven with an annuity for as long as he remained domiciled in Vienna; since it covered accidents and old age it also amounted to an insurance policy and a pension. The signatories were the Archduke Rudolph (1500 florins) and the Princes Lobkowitz (700 florins) and Kinsky (1800 florins). There were, as will be seen, difficulties in ensuring the regularity and the full value of the payments, but once those problems were overcome Beethoven was relieved from any rational grounds for financial worry.

Something must be said here about the Archduke Rudolph, the Emperor Franz's youngest brother, and not only the highest born but the most devoted of Beethoven's patrons. Born in 1788, he was destined for the church. As a boy he showed an aptitude for music, and at some time in his teens – perhaps in the winter of 1803–4, when he became 16 – he chose Beethoven as his piano teacher. Later he became Beethoven's only pupil in composition. The relationship, which lasted without interruption until Beethoven's death (Rudolph himself died four years later at the age of 43), was characterized by genuine respect on both sides. Rudolph treated Beethoven with consideration and humorous understanding; and Beethoven, though irked and sometimes provoked into ill-behaviour by the inevitable court protocol that surrounded a royal archduke, showed an almost childlike devotion to Rudolph, to whom he dedicated several of his greatest works. There are, it is true, many letters that show him begging off giving a lesson because of particularly pressing business or 'illness'; most of those pleas were accepted by the benevolent Rudolph as polite fictions.

The warmth of this relationship was to be highlighted by several incidents in the months that followed the signing of the annuity. For the second time within four years a French army bore down on Vienna, causing the imperial family, including Rudolph, to leave the city. Nevertheless it was decided that Vienna should be defended. As a result the city was bombarded by French howitzers throughout the night of 11 May and the following morning. Beethoven is said to have taken refuge in the cellar of Caspar Carl's house, and to have covered his head with pillows. On the afternoon of 12 May the city surrendered, and there was a second French occupation; it lasted for two months and proved a heavy drain on the inhabitants' pockets.

The summer of 1809 was a miserable one for Beethoven. Almost all his friends had, like the court, fled from the city, and communication with the outside world was greatly restricted. Nor could he search for inspiration and recreation in the countryside. He spent some weeks therefore in copying extracts from the theoretical works of C. P. E. Bach, Türk, Kirnberger, Fux and Albrechtsberger, as part of a course of instruction that he was preparing for the Archduke Rudolph. But his thoughts about his absent patron were expressed more touchingly in the programmatic 'Lebewohl' or 'Les adieux' Sonata (op.81*a*), the three movements of which depict his sorrowful farewell ('Das Lebewohl') to Rudolph on his departure from Vienna on 4 May 1809, his sadness at Rudolph's absence ('Abwesenheit'), and his rejoicing at seeing him again ('Wiedersehn') on his return on 30 January 1810. (Beethoven intended not only the titles but the dates to be inserted in the published work.) The sonata seems to have been completed in 1809 in anticipation of Rudolph's return, and was dedicated to him. Earlier in the year, before the French invasion, Beethoven finished the greater part of the Fifth Piano Concerto, also dedicated to Rudolph. The third important work of the year – like the concerto and the 'Lebewohl' Sonata in E♭ – was the so-called 'Harp' String Quartet (op.74). Several other smaller pieces were also completed before the end of the year: not only the F♯ major Piano Sonata (op.78), a work of which he himself thought very highly, but also the 'Sonatina' in G (op.79), the Piano Fantasia (op.77) and a number of songs. Beethoven's productivity even in one of his less productive years could be formidable.

Towards the end of the year a highly congenial commission came Beethoven's way, since it brought him in touch with the theatre once more, and since the play in question was by Goethe, whom he admired above all writers then living. It had been decided to furnish Goethe's *Egmont* with incidental music, and Beethoven was invited to supply it; he completed it by June 1810 and it was immediately performed. Apart from the excitement of the plot itself, in which Count Egmont foresees the liberation of the Netherlands from Spanish rule but dies as a result of his own brave stand, it is possible to suggest a deeper reason behind Beethoven's heartfelt response to it: it may represent his own delayed reaction to the conquest and occupation of his adopted city by the French, and his hopes of being delivered from them. In the spring or summer of 1810 he also wrote three songs (op.83) to words by Goethe, and he learnt about the poet's character through the friendship that he now formed with the very young, talented and seductive Bettina Brentano, a friend of Goethe – whom in turn she kept informed by letter about Beethoven.

Bettina obviously charmed Beethoven; rather less is known about another woman with whom he had been more seriously involved only a little earlier. For it seems clear that in the spring of 1810 Beethoven was more or less solemnly considering marriage. Not only did he turn his attention to his wardrobe and personal appearance; he even wrote to his old friend Wegeler in Bonn for a copy of his baptismal certificate, necessary evidence of his exact age. The woman who was the object of these concerns was a certain Therese Malfatti, the niece of Dr Johann Malfatti who had become his physician for a short while after the death of the trusted Dr Schmidt in 1808 (his doctor since about 1801). It looks as though Beethoven made a proposal of marriage and it was turned down. No doubt it was radically misconceived; there is no evidence that the family of Therese, who was not yet 20, would have found Beethoven, then in his 40th year, an acceptable suitor, and the one surviving letter from him to her, though warm enough, is not particularly intimate. Beethoven's disappointment is hard to gauge. He was urged to travel, perhaps because of his distracted state, but instead he merely moved to Baden for two months. The compositions on which he worked that summer include

51

the String Quartet in F minor (op.95) – the 'quartetto serioso' – and the so-called 'Archduke' Piano Trio in B♭ (op.97); although their autograph scores bear dates of October 1810 and March 1811 respectively, it is possible that both works were completed later than the dates suggest. The earlier months of 1811 seem to have been a time of comparative inactivity in composing, though a number of larger works, including the Choral Fantasy and the oratorio written several years earlier, had to be seen through the press.

Beethoven's health was still not satisfactory, and in the summer of 1811, on Dr Malfatti's orders, he visited the Bohemian spa Teplitz (now Tepliče) to take the cure. While there he wrote the incidental music to two stage works by Kotzebue, *König Stephan* (op.117) and *Die Ruinen von Athen* (op.113), designed as prologue and epilogue to the ceremonial opening of the new theatre at Pest. He evidently returned to Vienna refreshed and began work on the Seventh Symphony, which he completed in the spring of 1812, going on without a break to the Eighth Symphony. (To judge from the sketchbook used for work on these symphonies, he at one time considered following them with a third, probably in D minor.) For the second year running Beethoven decided to visit Teplitz, travelling via Prague and arriving there on 5 July. Next morning he started to write a love-letter to an unknown woman, which – since it has been discussed almost as much as any music he ever wrote – will be considered shortly. Because of the international situation (Napoleon's invasion of Russia was just getting under way), Teplitz, which was neutral territory, became the meeting-place of many imperial personages and diplomats. But what was even more interesting to Beethoven was the presence there of Goethe, and the long-awaited meeting between them finally took place. The contact was a cordial one, the reactions of the two men predictable. To his friend Zelter, Goethe confided:

His talent amazed me; unfortunately he is an utterly untamed personality, who is not altogether in the wrong in holding the world to be detestable but surely does not make it any the more enjoyable either for himself or for others by his attitude. He is easily excused, on the other hand, and much to be pitied, as his hearing is leaving him, which perhaps mars the musical part of his nature less than the social.

Beethoven's somewhat more censorious comment in a letter to Breitkopf & Härtel was: 'Goethe delights far too much in the court atmosphere, far more than is becoming in a poet'. In fact Beethoven's admiration for his fellow men usually flourished best at a distance.

From Teplitz he went, allegedly on a new doctor's advice, to Karlsbad (now Karlovy Váry), and from there to Franzensbrunn, where he participated in a charity concert held for the victims of a fire at Baden that had destroyed a large part of the resort. He then revisited Karlsbad, and finally returned once more to Teplitz, still apparently in search of improved health. At the beginning of October he was in Linz, where he started the score of the Eighth Symphony; he stayed with his brother Johann, who had bought an apothecary's shop there in 1808. But this was less of a visit than a visitation, for the principal purpose of his journey to Linz was to interfere in his brother's private life. Johann had let part of his house to a physician from Vienna, whose wife's unmarried sister, one Therese Obermeyer, later joined them. Subsequently Therese became Johann's mistress, and Beethoven now descended to expostulate with his brother and to attempt to end the relationship. He applied both to the bishop and to the civil authorities, and ultimately obtained a police order to have the girl expelled from Linz. But before it could be effective Johann played a trump card by marrying Therese, on 8 November. Beethoven's extravagantly high-handed behaviour had ended in defeat, and he retired angrily to Vienna. Nothing more is heard of him that year apart from the preparations for a concert with the French violinist Pierre Rode on 29 December, for which he completed the G major Violin Sonata (op.96).

The rebuff by his brother was the second emotional crisis of 1812, a year that represented some sort of watershed for Beethoven. To return to the letter of 6–7 July: usually known as the letter to the 'Immortal Beloved' ('unsterbliche Geliebte'), it was found among Beethoven's papers after his death, and first published in 1840. There is no direct indication to whom this passionate love-letter, the only one of his to a woman that uses the intimate 'du' throughout, was addressed. Even the year in which it was written (it refers only to 'Monday, 6 July') was for long uncertain. Thus the names of many women known to have been admired by Beethoven were proposed by his early biographers; but nearly all of them have had to be ruled out, since 1812 is now established as the correct date of the letter, Teplitz as its place of origin and Karlsbad ('K' in the letter) as its addressee's temporary residence.

Of recent conjectures as to her identity the most plausible (by Maynard Solomon) is that she was Antonie Brentano, an aristocratic Viennese lady ten years younger than Beethoven who at 18 had married a Frankfurt businessman, Franz Brentano, Bettina Brentano's half-brother. The Brentanos were in Vienna in the years 1809–12, so that Antonie could be with her dying father and subsequently wind up his estate. It is clear not merely that she disliked the idea of returning to Frankfurt, where she was most unhappy, but that she did everything possible to postpone it, delaying the event until the last months of 1812. Beethoven had been introduced to the family by Bettina in 1810, and became a warm friend not only of Antonie but of her husband Franz and their ten-year-old daughter Maximiliane – for whom in June 1812 he wrote an easy piano trio in one movement (WoO 39). Since the Brentanos had not only been in close contact with Beethoven in Vienna shortly before his departure at the end of June, but were also in Prague while he was there (2–4 July) and moved on to Karlsbad on 5 July, Antonie Brentano fulfils all the chronological and topographical requirements for being the addressee of the famous letter.

Whether the psychological requirements are fulfilled depends on one's reading of her personality and of the letter's inner meaning. Although in many ways a dutiful wife, her admiration of Beethoven was profound, and she may have become emotionally dependent on him, especially when the return to Frankfurt seemed inevitable. And there is no doubt that Beethoven, though vociferous in his condemnation of adulterous relations, was especially attracted to women who were married or who were in some other way already involved with a man. In any case the letter is a highly ambiguous document. Mingled with the ardently expressed desire for complete union with the beloved ('I will arrange it with you and me that I can live with you') there are many phrases expressing resignation or acceptance of the lack of fulfilment, and it is possible to read it as a cautious rejection of a shared domesticity: 'At my age I need a steady, quiet life – can it be so in our connection?'. Doubtless the ambiguities were clarified when, later in the month, Beethoven joined the Brentanos at Karlsbad. The family duly returned to Frankfurt in the autumn; Beethoven never saw them again, though he remained in touch with them, calling on Franz's services as a businessman in 1820 and dedicating important works to Antonie (op.120, in 1823) and Maximiliane (op.109, in 1821).

VIII 1813–21

However the turmoil of the summer of 1812 is to be understood, it proved to be a profound turning-point in Beethoven's emotional life. It initiated a long period of markedly reduced creativity, and there is evidence that he became deeply depressed. Henceforth Beethoven accepted the impossibility of achieving a sustained relationship with a woman and entering into a shared domestic routine, though he was scarcely reconciled to it; even in 1816, as will be seen, he had by no means overcome his longing. Some of the hints contained in the letter are stated more baldly in diary entries made about this time. As in past crises, a dedication to art was evidently to replace a commitment to a human being: 'Thou mayst no longer be a man, not for thyself, only for others, for thee there is no longer happiness except in thyself, in thy art – O God, give me strength to conquer myself, nothing must fetter me to life'. '13 May 1813. To forgo a great act which might be and might remain so ... O God, God, look down on the unhappy B., do not let it continue like this any longer.'

But by a stroke of irony that may contain an inner truth, at this very time he pledged himself to a responsibility that was increasingly to encroach on the exercise of his art and indeed to dominate his emotional outlook in the last 12 years of his life. Caspar Carl became seriously ill with tuberculosis, and on 12 April 1813 he signed a declaration appointing Beethoven guardian of his son Karl, then aged six, in the event of his death. This, it will emerge, came ultimately to involve Beethoven profoundly; but for the moment Caspar Carl's health improved, though Beethoven was obliged to help him to borrow money.

At this time Beethoven too was financially embarrassed. The severe depreciation of the Austrian currency as a result of the war, leading to an official devaluation in February 1811, had reduced the value of his annuity of 4000 florins to little more than 1600 florins. It was open to the princes to restore the intended income, and they were prepared to do so; but unfortunately Prince Kinsky was killed by a fall from his horse at the end of 1812 before he could leave clear instructions, and Prince Lobkowitz's payments were suspended for four years from 1811 owing to the mismanagement of his affairs. So although Beethoven was ultimately to receive the full amount from Kinsky's heirs, from Lobkowitz and from the Archduke Rudolph, it was only the last-named whose subventions continued without interruption or depreciation.

This may be one reason why Beethoven, even though he was still nursing secret sorrows, nevertheless became more of a public and social figure in the next year or so, reaching for popular acclaim by way of the concert hall and the theatre. He not only engaged a servant, but appears to have kept him for three years. And he entered with some zest into the proposal of Johann Nepomuk Maelzel, the inventor of a mechanical organ called the 'panharmonicon' (and later, inventor of a metronome), for the two of them to collaborate on a piece that both celebrated and depicted Wellington's military victory at Vittoria on 21 June 1813. This absurd piece of programme music, with its fanfares, cannonades, and fugal treatment of *God Save the King*, was thunderously acclaimed at two charity concerts on 8 and 12 December 1813 – together with the Seventh Symphony, which had not been heard before. The 'Battle Symphony' had to be repeated three weeks later, and again on 24 February 1814. On that occasion the Eighth Symphony was one of its companion pieces.

The most gratifying (and unexpected) consequence of this sudden popularity was a request from the Kärntnertor Theater for permission to revive the opera *Fidelio*. Beethoven agreed but stipulated that there would have to be a good many changes. The poet

4. Ludwig van Beethoven: bronze bust (1812) by Franz Klein

G. F. Treitschke was then stage manager at the theatre, and he undertook to make the necessary alterations in the libretto. Some weak numbers were omitted, the two finales were rewritten, Leonore's aria was supplied with a new recitative ('Abscheulicher!') and Florestan's with a new final section, and there were many smaller changes throughout. Beethoven also furnished the revival with a new overture in E major, called today the 'Overture to *Fidelio*'. Although he grumbled at the labour and claimed in a letter to Treitschke that the opera would win him a martyr's crown, the revision was effective, and the work's success dates from this production, first given on 23 May 1814. The new overture, not ready for the opening night, was given at the second performance, on 26 May.

The vocal score of the opera was prepared by the young pianist Ignaz Moscheles, then just 20. Since he worked under Beethoven's supervision, the task brought him for a time into regular contact with someone he had for long ardently admired. And March 1814 was the date at which another enthusiastic follower of Beethoven later said he had first been introduced to him: this was the 18-year-old Anton Schindler, at the time a law student and a good violinist. For Schindler the claims of music proved stronger than those of the law, and by 1822 he was leader of the orchestra at the Josephstadt Theater. From about that time he began to spend many of his leisure hours in Beethoven's company, and for a while he virtually became his unpaid secretary and servant. Beethoven found his 'factotum' useful in practical matters, though Schindler's obsequiousness used to irritate him. Some years after Beethoven's death, in 1840, Schindler published a hastily written biography – translated into English a year later, with notes by Moscheles – in which uncritical devotion to 'the master' was combined with polemics against many of the others who had been close to him. Thus it is unfortunately unreliable even in its account of the years after 1821 during which Schindler was often in very close contact with the composer; the material of value that it contains is hard to distinguish from his fabrications. A later (1860) edition of his biography, although greatly expanded and indeed largely rewritten, was no more accurate.

In the summer of 1814 excitement began to mount in Vienna as preparations were made to welcome the crowned heads of Europe for the Congress of Vienna. This gave Beethoven the opportunity for producing more 'occasional pieces'. But before starting work on anything of that nature, he quickly completed a piano sonata (op.90), his first in four years. The earliest of the congress works was a short chorus of welcome to the visiting sovereigns, *Ihr weisen Gründer* (WoO 95). Next, he made a strenuous attempt to complete an overture in C that he had taken up and worked on at various times in the previous five years; it was planned for the celebration of the emperor's name day on 4 October and is now known as the *Namensfeier* Overture (op.115). But the score could not be completed in time, and Beethoven put it aside until the spring of 1815, setting to work instead on a cantata celebrating the present 'glorious moment' in the destiny of Europe. The absurdly bombastic text of *Der glorreiche Augen-*

blick (op.136) was by a distinguished surgeon from Salzburg, Alois Weissenbach, who had come to the capital for the festivities. Beethoven could not have had a more enthusiastic admirer than Weissenbach; when the two men met, they took a great liking to each other, and the cantata was a result of their collaboration. They had more than music in common, for Weissenbach too was deaf. The cantata was announced for a concert on 20 November, but it was postponed three times and finally given before the assembled royalty on 29 November, with the 'Battle Symphony' and Seventh Symphony forming the rest of the programme.

From the point of view of Viennese popular acclaim and fame the year 1814 must be regarded as the high-water mark in Beethoven's life. Not only were his compositions applauded by large audiences, but he also received in person the commendations of royal dignitaries. This last aspect is typified in one final congress piece, the little Polonaise (op.89) that he wrote in December 1814 in honour of the Empress of Russia, who was especially generous to him. And 1814 was also a more sombre turning-point for Beethoven, for two performances of the 'Archduke' Trio in April and May marked his last appearance in public as a pianist (except as accompanist). His deafness had latterly become much more severe.

Beethoven now found himself possessed not only of fame but of a good deal of money, which he invested in bank shares. Moreover, as a result of a settlement reached with the Kinsky family and the goodwill of Prince Lobkowitz, most of the original value of the annuity had now been restored and the arrears made up. In spite of this, his worries about his financial situation continued to be voiced in letters to publishers and friends abroad (such as his former pupil Ries, now resident in London), whom he was trying to interest in the large number of his more recent works that were still unpublished. But towards the end of 1815 an unhappy event occurred that immediately focussed all his concerns and anxieties. His brother Caspar Carl's health suddenly deteriorated; the tuberculosis had evidently made inconspicuous but rapid progress, and he collapsed and died on 15 November. The will, dated 14 November, appointed Beethoven sole guardian of his only child, the nine-year-old Karl, but a codicil of the same day cancelled this and made the boy's mother co-guardian:

Having learnt that my brother ... desires after my death to take wholly to himself my son Karl, and wholly to withdraw him from the supervision and training of his mother, and inasmuch as the best of harmony does not exist between my brother and my wife, I have found it necessary to add to my will that I by no means desire that my son be taken away from his mother, but that he shall always and so long as his future career permits remain with his mother, to which end the guardianship of him is to be exercised by her as well as by my brother ... for the welfare of my child I recommend *compliance* to my wife and more *moderation* to my brother. God permit them to be harmonious for the sake of my child's welfare. This is the last wish of the dying husband and brother.

The dying man's anxieties were all too prophetic. It proved a tragedy for Beethoven that he could not do what his brother asked. It was not simply that he was unable to achieve any 'harmony' with his sister-in-law Johanna. The situation in which he found himself was one that aroused deep passions and longings that he doubtless did not fully understand. Frustrated in his several attempts – however ambiguously conceived and executed – to marry and have a family of his own, he began to feel that if he had sole responsibility for Karl he could combine the discharge of a sacred duty to his brother with some of the satisfactions and fulfilments of parenthood. But for that to be possible, he had first to convince himself and others that Johanna was quite unfit to have the custody of Karl and should be excluded from the guardianship. The struggle for possession of the nephew lasted some four and a half years, to be followed by another six in which his care and upbringing weighed heavily upon Beethoven. As will be seen, the burdensome intensity of the relationship between uncle and nephew – or as Beethoven preferred to see it, between father and son – led to something near disaster in the summer of 1826. Before then an incalculable number of hours had been spent by Beethoven in litigation, letter-writing, quarrels, reconciliations and private agony of mind.

On 22 November 1815 the Imperial and Royal Landrechte of Lower Austria appointed Johanna guardian and Beethoven 'co-guardian'. Six days later Beethoven appealed to the court requesting the guardianship to be transferred to himself; in a later court appearance he claimed he could produce 'weighty reasons' for the total exclusion of the widow from the guardianship. The 'weighty reasons' consisted of evidence that four years earlier Johanna had been sentenced to a month's house arrest on a charge of embezzlement brought by her husband. The result of Beethoven's submissions was that on

9 January 1816 he was assigned sole guardianship by the court. He took vows for the performance of his duties on 19 January. On 2 February Karl was taken from his mother and entered the private school of a certain Cajetan Giannatasio del Rio as a boarder.

Beethoven seems to have had little difficulty in persuading himself that Johanna was morally quite unfit to have charge of Karl, and he was ready to denounce her character and her way of life on every possible occasion, calling her the 'Queen of Night' and insinuating that her allegedly deviant behaviour included prostitution and theft. She was certainly no moral exemplar, and some time after her husband's death she took a lover and gave birth in 1820 to an illegitimate daughter. But Viennese society was permissive in sexual matters. Few of her contemporaries saw her in the same lurid light as her brother-in-law, in spite of the forceful and relentless way that he marshalled the case against her.

Although convinced of Johanna's unsuitability for bringing up the child, Beethoven felt rather guilty about restricting them from seeing each other. Yet that is what he now asked the Landrechte to put in his control, and the court agreed that Johanna should visit her son only at hours and places that Beethoven sanctioned – which at times was liable to mean once a month, or even less frequently. An uneasy truce was maintained between Beethoven and Johanna through 1816 and 1817, although he suspected her of making clandestine visits to Karl's school. At the end of January 1818 he withdrew Karl from Giannatasio's care and took him into his own home, engaging a private tutor; then in May he moved with Karl to Mödling and placed him in a class taught by the village priest, named Fröhlich. But after a month, to Beethoven's indignation, Fröhlich expelled Karl for his bad behaviour. This seems to have consisted of a series of minor offences against discipline, but Karl particularly shocked the priest by speaking of his mother in abusive terms – a breach of the Fifth Commandment in which, it was later noted, Beethoven had gleefully encouraged him.

It was at this point in the summer of 1818, when Beethoven was taking preparatory steps to enter Karl in the Vienna Gymnasium, that Johanna made a further effort to gain some control of her son's education and welfare. With the help of a relative with legal training, Jacob Hotschevar, she presented a series of petitions to the Landrechte. The first two were rejected, but after Karl had run away from Beethoven's lodgings to his mother on 3 December – he was returned later by the police – she used the incident as the basis of a third appeal, supporting it by a careful summary of the whole situation from Hotschevar and appending a statement from Father Fröhlich on the boy's neglected physical state and moral lapses. In the course of giving evidence in court on 11 December Beethoven incautiously let slip the fact that Karl was not of noble birth. He was then forced to concede that neither he nor his late brother had ever had documents to prove their own nobility; 'van' was a Dutch prefix that was not restricted to those of noble birth. Thereupon the Landrechte, which were courts confined to the nobility, woke up to the fact that the case should never have come before them and transferred the whole matter to the Vienna Magistracy or commoners' courts.

How severe a blow this was to Beethoven's pride has been debated; but even from a practical point of view it was very inconvenient. From the start the Magistracy seems to have been more sympathetic to Johanna than to Beethoven. Its first action was to suspend him temporarily from the guardianship. Karl returned for a time to his mother, being instructed by a tutor and also being taught at an institute run by one Johann Kudlich. From March to July Beethoven resigned the guardianship in favour of a Councillor Tuscher, and applied for a passport to enable Karl to be educated in Bavaria. This was refused, and his right to resume the guardianship in July was also challenged; on 17 September the court decided, reasonably enough, that Karl (who had meanwhile been moved to yet another school, one run by a Pestalozzi disciple, Joseph Blöchlinger) had been 'tossed back and forth like a ball from one educational institution to another'. The mother, therefore, should remain as legal guardian in collaboration with a certain Leopold Nussböck, the municipal sequestrator.

This was of course a defeat for Beethoven. His first move was to protest at the decision; this was rejected by the Magistracy on 4 November. Next, with the help of a legally qualified friend, Johann Baptist Bach, he proposed as a substitute for Nussböck his friend Karl Peters, who was tutor to the children of Prince Lobkowitz. This application too was rejected. He now had recourse to the Court of Appeal, for whose benefit he prepared a 48-page draft memorandum (the longest extant document in his handwriting). This denounced in turn Johanna, a certain Herr Piuk who was a member of the Magistracy, and Father Fröhlich, and defended his own conduct and educational policies in great detail. It is unlikely that the memorandum was ever submitted in the form in which it survives. Beethoven's case was shrewdly and discreetly presented by Dr Bach; Beethoven persisted in demanding the guardianship of Karl and requested Karl Peters as associate guardian, asking at the same time for Johanna and Nussböck to be deposed. After further scrutiny these claims were upheld by the Court of Appeal on 8 April 1820; a petition by Johanna to the emperor against the decision was rejected three months later. Thus in July 1820 Beethoven found that he had finally won in a struggle that had lasted for over four years.

Beethoven's preoccupation with the care of his nephew – especially in the period from the end of 1815 to the beginning of 1818 – can be regarded as a continuation, and in some ways as an attempted solution, of the unresolved matrimonial crisis of 1812. At that time he had decided, however confusedly and irresolutely, that his creative activity was incompatible with having a wife; now he was testing whether it could be reconciled with caring for a child. The cost of those years to Beethoven is reflected in the paucity of valuable music completed in them. Productively the years 1813–15 were lean; apart from two cello sonatas (op.102) written in the second half of 1815, most of the compositions were 'occasional pieces' such as the 'Battle Symphony' and the works written for the

Congress of Vienna. This trend was continued in the following years. 1816 at least brought two important compositions, the song cycle *An die ferne Geliebte* (op.98, April) and the Piano Sonata in A (op.101, November), but 1817 was completely barren in respect of completed major works. Instead, Beethoven during these years contented himself with elegant trivia, such as the polished march that he wrote in June 1816 for the Vienna artillery corps (WoO 24), and the instrumental and vocal settings of Scottish airs that he provided for George Thomson of Edinburgh (opp.105, 107, 108 etc). He also refurbished some variations for piano trio that he had written many years earlier on a theme from one of Wenzel Müller's Singspiels, and he revised and to some extent rewrote an arrangement for string quintet made by someone else of his youthful C minor trio from op.1 (these were subsequently published as his op.121*a* and his op.104). And, even more significantly, he toiled hard on a number of new compositions without managing to complete them; they included a piano concerto in D, a piano trio in F minor, and a string quintet in D minor. Scores of these three works were in fact begun.

These were indeed unhappy years for Beethoven. He was now thoroughly out of sympathy with the kind of music being written and being applauded in Vienna. The aristocratic milieu that had welcomed and sheltered him in his earlier years in Vienna had been shattered by the military, political and financial upheavals of the Napoleonic wars, with the result that he had lost or broken with almost all his high-born friends apart from the Archduke Rudolph. In spite of his popular successes in 1813 and 1814 and his general acceptance as the greatest living composer, he found no wide public in Vienna that he could respect, and daydreams of journeys abroad – to England, even to Italy – filled his mind. Nor should it be supposed that the attachment to Antonie Brentano, though he had not seen her for some years, was forgotten. The best informant here is Fanny, one of Giannatasio's daughters, who observed Beethoven at this time with a sensitivity sharpened by her own unavowed devotion to him. In September 1816 she recorded in her diary a confession of Beethoven to her father that she had overheard. Five years before, he had wanted a more intimate union with a woman, but it was 'not to be thought of, almost impossible, a chimera. Nevertheless, it is now as on the first day, I have not been able to get it out of my mind'. Some months earlier, on 8 May 1816, Beethoven had ended a letter to Ries in London with the words: 'My best greetings to your wife. Unfortunately, I have no wife. I found only one whom I shall doubtless never possess'. This nostalgic retrospection forms the background to the song cycle on the subject of the 'distant beloved' that he wrote in April 1816.

There were also difficulties of a more practical kind. Beethoven was consumed with misgivings as to his ability to look after his nephew and to run an orderly household. The year 1817, in particular, is marked by an immense number of letters to the kindly Nannette Streicher, a pianist and wife of the piano maker Johann Andreas Streicher, on the minutiae of domestic administration, the cost of household commodities, the employment of servants, and the like. Deeper doubts about the decisions that he was taking on Karl's behalf and about his treatment of Johanna were committed to his diary:

God, God, my refuge, my rock, my all. Thou seest my inmost heart and knowest how it pains me to be obliged to compel another to suffer by my good labours for my precious Karl!!! O hear me always, thou Ineffable One, hear me – thy unhappy, most unhappy of all mortals.

Further problems were created by his slowly but unmistakably deteriorating health and especially by one aspect of it, his deepening deafness. By 1818 he was virtually stone deaf, so conversation had to be carried on with pencil and paper. This was the start of the 'conversation books', nearly 140 of which have survived. In the main they are a record of only one side of each discussion; they show what Beethoven's friends and visitors wanted to say to him, but not his own observations, since those were normally spoken. Unfortunately Schindler, who took possession of the conversation books after Beethoven's death, saw fit not only to destroy some of them but to make false entries in the remainder, so that as documents they must be treated with some caution.

Beethoven's recovery from his compositional stagnation seems to have begun in the autumn of 1817. It was at first very slow. At that time he decided to accept an offer made earlier in the year by the Philharmonic Society of London. This invited him to write two grand symphonies for the Society, and to appear in person in London for the winter season of 1817–18. But he made no start on a symphony, or plans for a journey to

London, afterwards explaining that his health had not allowed it. Instead, he set to work on a gigantic four-movement piano sonata in B♭, known today as the 'Hammerklavier' Sonata (op.106). Its first two movements were probably ready by April 1818, and the remaining two were worked on during his summer stay at Mödling, the whole being completed by the autumn. Thus its composition, carried out (as he said) 'in distressful circumstances', had taken the best part of a year. Beethoven dedicated it to the Archduke Rudolph, for whom he was now planning a work on an even grander scale. For the archduke was being made the recipient of ecclesiastical honours. He was created a cardinal on 24 April 1819, and on 4 June he was appointed Archbishop of Olmütz (now Olomouc) in Moravia. 'The day', wrote Beethoven in offering his congratulations on the latter elevation, 'on which a High Mass composed by me will be performed during the ceremonies solemnized for Your Imperial Highness will be the most glorious day of my life', and it looks as though by then he had already been at work for some time on the composition now known as the *Missa solemnis* (op.123). Evidently the news that the archduke was to be elevated had been known to friends in advance.

Since the installation of the archbishop was set for 9 March 1820, some way ahead, Beethoven must have felt that he could afford to proceed at a measured pace. He even interrupted work on the Mass to write down some 20 variations on a theme of Anton Diabelli's before tackling the Gloria and Credo. But he had not allowed for the time about to be lost in litigation in

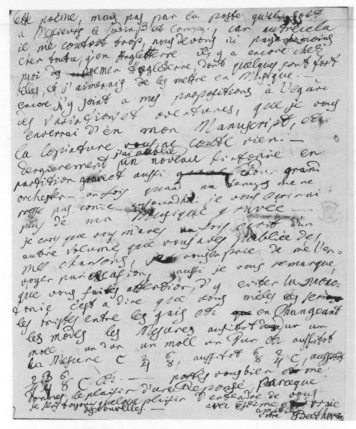

5. *Part of an autograph letter (21 February 1818) from Beethoven to George Thomson of Edinburgh*

1819 and the first months of 1820, or for the tendency of each section of the work to expand to a vast scale. Beethoven had to abandon any hope of the mass's being ready for the installation. But he persevered with it, and even took on new commitments, undertaking at the end of May 1820 to produce three piano sonatas within three months for the Berlin publisher Adolf Martin Schlesinger. Although nothing like that optimistic pace was achieved, the first sonata was apparently completed and a start made on the other two shortly after his return to Vienna from Mödling in the autumn of 1820. The sonata that was now ready was the one in E, published as op.109. But in 1821 illnesses both at the start of the year and in July – this time an attack of jaundice – as well as the work on the mass resulted in the other two sonatas not coming near to completion until the end of the year. The autograph of the second, in A♭, is dated 25 December 1821, that of the third, in C minor, 13 January 1822; but revisions to both postponed their completion for a little longer. Unlike op.109, published in Berlin, the other two (opp.110 and 111) first appeared in Paris from the firm that Adolf Martin Schlesinger's son Maurice had started there.

IX   1822–4

There was no longer any question of checks on Beethoven's creativity. 1822 saw not only the finishing touches on the two sonatas, the last he was to write, but the virtual completion of the Mass by the autumn, and an almost immediate start on another very large composition that he was impatient to get to grips with. This was the work now known as the Ninth Symphony. Before that he had also assembled a set of 11 bagatelles for the piano (op.119), five of which had been written by the beginning of 1821 for an instructional book of studies (most of the others were based on much earlier material); and he resumed work on the set of piano variations on Diabelli's theme that he had broken off in 1819. He found time, too, to compose a fine overture (op.124) and a chorus (WoO 98) for the opening of the new theatre in the Josephstadt on 3 October 1822. The overture, *Die Weihe des Hauses* ('The Consecration of the House'), takes its title from the inaugural drama at the theatre.

The piano variations need a word of explanation. In 1819 Diabelli, no doubt responding to post-Congress patriotic fervour and in search of attractive publishing material for the new firm of Cappi & Diabelli, conceived the idea of inviting a large number of eminent or popular composers from the Austrian states to submit a single variation on a simple theme of his own that he circulated; the intention was to make an album. Such an album was indeed published by Diabelli, though not until 1824, with variations from 50 composers including Schubert and the 11-year-old Liszt. But from the start Beethoven had decided to contribute not one variation but a set of them. In time these reached the number of 33, and Diabelli decided to publish Beethoven's variations (op.120) as a separate album; in fact it came out before the other one.

The nature of the symphony to which Beethoven now turned his attention can be understood as the coalescence of several diverse elements that had been stirring in his imagination, in some cases over many years. The notion of composing a vocal setting of Schiller's *An die Freude* ('Ode to Joy') goes back to his last days in Bonn, as a letter of January 1793, from the Bonn professor of jurisprudence Fischenich to

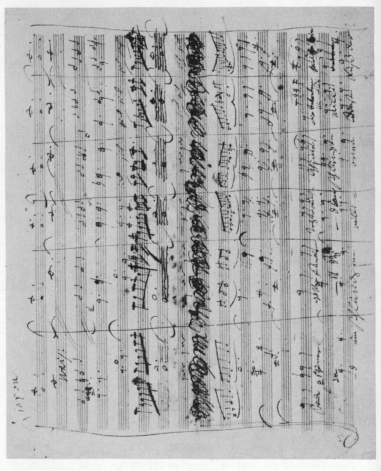

6. Autograph MS
of part of the last
movement of
Symphony no.9,
composed 1821–4

76

Schiller's wife, makes clear: 'He proposes to compose Schiller's *Freude*, strophe by strophe. I expect something perfect, since he is wholly devoted to the great and sublime'. This was an intention to which he returned a number of times – in 1798 for instance, and in 1812, in connection with sketches for an overture that later became the *Namensfeier*. Another element was the desire to complete at least one symphony for the Philharmonic Society, and possibly the promised two. For a time it seems that he conceived of one of these symphonies as containing a choral section – a 'pious song in a symphony in the ancient modes' – and the other as being in D minor without any such special feature. Only in 1822 were these diverse concepts united in the plan for a D minor symphony with a setting of Schiller's *Ode* as its finale: this he now intended to conclude with Turkish music and a full choir.

1823 was the year in which the main work on the Ninth Symphony was done, though the last details were not completed until the following March. It was also a year of great concern with copyists and publishers. Beethoven made the mistake of offering manuscript copies of his mass on a subscription basis – at a price of 50 ducats – to the crowned heads of Europe; this involved him first in a tedious correspondence with the courts, and then in a no less irksome scrutiny of the handwritten scores (a task for which Schindler was put to use). The difficulties were increased by the illness and death on 6 August 1823 of Wenzel Schlemmer, who had been Beethoven's chief copyist for a quarter of a century and on whom he relied greatly. This year also

77

saw the publication of the op.111 piano sonata and the
'Diabelli' Variations. The mass formed the centre of
an immensely complicated series of negotiations with
publishers in Vienna and abroad, in which other works
completed and uncompleted, such as the op.124 overture and the Ninth Symphony, also featured. It must be
remembered that he regarded the mass as his greatest
work, the result of some two years' labour and not
lightly to be disposed of; if anything was outrageous it
was not the size of the fee demanded but the fact that by
the end no fewer than seven publishers had been
involved. The final result was satisfactory: a firm that
he could trust, Schott of Mainz, agreed to publish
several of his important works, including the mass
and the Ninth Symphony.

With the symphony completed Beethoven allowed
himself some relaxation; according to Schindler, 'he
could again be seen strolling through the streets, using
his black-ribboned lorgnette to examine attractive
window displays, and greeting many acquaintances or
friends after his long seclusion'. But as he was now
thoroughly out of sympathy with the musical taste of
Vienna, he was reluctant to risk a concert, and made an
inquiry of Berlin whether a performance of the mass
and the symphony might be given there. News of this
fact became known in Vienna and led to a touching
document being presented to him by a number of his
friends and admirers. This was an eloquent declaration
of their confidence in him, and a plea for him to allow
his latest works to be heard in Vienna. Beethoven
responded by agreeing to give a concert. It took place in
the Kärntnertor Theater on 7 May 1824 and consisted

of the op.124 overture, the Kyrie, Credo and Agnus
Dei from the mass, and the Ninth Symphony. The
theatre was crowded and the reception enthusiastic.
Many years later the pianist Thalberg, who was among
those present, recalled that after the scherzo had ended
Beethoven stood turning over the leaves of the score,
quite unaware of the thunderous applause, until the
contralto Caroline Unger pulled him by the sleeve and
pointed to the audience behind him, to whom he then
turned and bowed (Schindler and Mme Unger also
remembered the moving incident, though they placed it
at the end of the concert). A second performance of the
symphony and the Kyrie of the mass (with some other
pieces) 16 days later was much less successful.

Around the time of the symphony's first performance in May 1824, Beethoven turned once more to
the piano and wrote a 'cycle of bagatelles'. Unlike the
earlier ones he had written (opp.33, 119), the six
bagatelles of op.126 were conceived not as separate
pieces but as forming a set. At the end of the year he
returned to a poem that he had come to value highly.
This was Matthisson's *Opferlied*, which he seems to
have regarded (in Nottebohm's phrase) as 'a prayer for
all seasons'. He had set it in 1795 and again in 1822;
now he produced his final version, a setting for
soprano, chorus and orchestra (op.121*b*). In these
years, when Beethoven was hoping that his smaller
pieces at any rate would prove easy to sell, he was no
doubt tempted to refurbish drafts of songs written
many years earlier and to put them on the market. But
with the *Opferlied*, as with the much better-known
instance of Schiller's *Ode to Joy*, one may detect some

79

elements of a desire in Beethoven around this time to gather up the unfinished business of the past and attend to ideas that had waited long for definitive expression. He was already beginning to suspect that not much time was left to him.

X 1824–7

It seems unlikely that anyone could have predicted that the remaining years of Beethoven's life would be devoted to works in a single medium – that of the string quartet. Since 1810 he had composed no quartets. In the miraculously fertile year of 1822, however, he had written to the publisher Peters on 5 June quoting his price (50 ducats) for a string quartet 'which you could have very soon'. A letter of a month later explained that the quartet was 'not yet quite finished, because something else intervened'. It is unlikely, however, that by then he had even started to work on the Quartet in E♭ (op.127). The impetus to complete it and to compose others was provided by a commission from Prince Nikolai Golitsïn (Galitzin), a music lover and cellist of St Petersburg. In a letter of 9 November 1822 Golitsïn invited Beethoven to compose 'one, two or three new quartets' for whatever fee was thought proper; they were to be dedicated to the prince. In his reply of 25 January 1823 Beethoven accepted the invitation, fixing his honorarium at 50 ducats per quartet and promising to complete the first by the end of February or by the middle of March at latest. But he had not allowed for the claims of the mass and the symphony; not until after the concerts of May 1824 was the work resumed in earnest. The quartet was finished in February 1825, nearly two years after it had been promised, and was privately rehearsed before being sent to Golitsïn. In the meantime Golitsïn, who had been among the princely subscribers to the manuscript copies of the mass that Beethoven had advertised in 1823, gave the first performance of that work at St Petersburg on 7 April (26 March, Old Style) 1824 – a whole month before the partial performance in Vienna.

The E♭ Quartet was performed by the Schuppanzigh Quartet on 6 March 1825, but did not please the Viennese audience. Later performances, however, in which Joseph Boehm led instead of Schuppanzigh, were well received. Beethoven at once turned his attention to the second quartet for Golitsïn, in A minor (op.132). Some progress had already been made when a sharp illness in April sent him to his bed. He was ill for about a month, but felt well enough by 7 May to move to Baden, and there the quartet was completed in July. Its slow movement contains allusions to his illness; the initial melody is inscribed 'Hymn of thanksgiving to the divinity, from a convalescent, in the Lydian mode', and the contrasting section in 3/8 time is entitled 'Feeling new strength'. This work received two private performances from the Schuppanzigh Quartet on 9 and 11 September 1825; among the audience was the publisher Maurice Schlesinger, who agreed to buy it, as well as another quartet not yet written, at the price of 80 ducats per quartet (the one in E♭ had already gone to Schott). The first public performance of the A minor Quartet was on 6 November, again by the Schuppanzigh Quartet.

81

Without any break Beethoven started work on Golitsïn's third quartet, which occupied him from July to December 1825. The Schuppanzigh Quartet gave its première on 21 March 1826. This work, in B♭ (op.130), consisted of six movements, the last of which, an immense fugue, proved something of a stumbling-block to players and listeners. No doubt this work too should have gone to Schlesinger to publish, but in the end Beethoven gave it to the Viennese firm of Matthias Artaria.

Beethoven had now fulfilled his commission, but Prince Golitsïn had paid only for the first of his three quartets; he still owed Beethoven 125 ducats – 50 ducats for each of the other two quartets, and 25 ducats for the dedication of the op.124 overture. Although the prince acknowledged the debt, and expressed himself immensely pleased with the quartets, he was financially embarrassed at the time, and his promise to pay was not carried out before Beethoven's death.

By the beginning of 1826, if no earlier, Beethoven was at work on a fourth quartet, in C♯ minor (op.131). Just as in the previous year, while he had been engaged on the A minor quartet, so now illness once again interrupted him. As before it was abdominal pain, and seemingly pain in his joints; his eyes were also affected. But before the end of March he was better, and completed the quartet in all essentials by June. This quartet was plainly intended for Maurice Schlesinger, to whom he had written on 22 April with a request for 80 ducats straight away, 'for quartets are now in demand everywhere, and it really seems that our age is taking a step forward'. But Schlesinger's Paris firm had been damaged by fire, and on getting no reply Beethoven impatiently offered the quartet to Schlesinger's father in Berlin, to Probst of Leipzig, and to Schott of Mainz, who secured the work.

To understand the events of the summer of 1826 it is necessary to go back some way and resume the story of the nephew at the point that it was broken off in 1820. After the guardianship issue had been resolved in Beethoven's favour in that year, Karl remained at Blöchlinger's educational institute until the summer of 1823. Having by then matriculated, he proceeded to the university and attended the philological lectures that were given there. He was just 17, and in spite of the earlier forebodings of Beethoven and Blöchlinger about his character and his industry, the almost complete segregation from his mother that he had to endure, and the conflicts of loyalty constantly imposed on him, he had developed well and had shown good progress in his studies. He was also making himself useful to his uncle, with whom he spent the summer of 1823 in Baden, acting as messenger and handyman, and sometimes as amanuensis and ready-reckoner. When Beethoven returned to Vienna for the winter Karl moved in with him, and remained until Easter 1825, when he left the university for the Polytechnic Institute and moved to lodgings run by a certain Matthias Schlemmer.

Whether they were living together or apart, it was not an easy relationship. From the conversation-book entries Karl appears as good-natured, lively and shrewd, but perhaps also a little sly and prone to tell tales; he must after all have been used to hearing people slandered recklessly, and he was eager to please his intimidating uncle. Beethoven's helplessness in practical matters, which included dealing with the servants, put a heavy load on Karl's time; but his possessiveness, suspiciousness and irritability must have been even more of a burden. Beethoven was jealous of Karl's young friends, and not only disparaged them but tried to prevent him from seeing them; at the same time, when he had moved for the summer to Baden, he expected Karl to come out to visit him on Sundays and holidays, thus greatly interfering with his nephew's studies.

In 1825 Beethoven himself acquired a friend nearer to Karl's age than to his own. This was Karl Holz, the second violin in the Schuppanzigh Quartet, who was then 27. Holz came to occupy something of the same place in his household that had previously been held by Schindler; Schindler was more or less completely displaced by Holz during 1825 and most of 1826, and never forgave him. The conversation-book entries suggest that Beethoven began to use Holz to spy on Karl.

The letters of Beethoven to Karl in the years 1825 and 1826 are full of reproaches and recriminations, and demands for his affection and attention. There are also violently emotional attempts at reconciliation. The conversation books tell the same story: Beethoven was ceaselessly suspicious of the friends Karl had, the use he made of his spare time, the way he spent his money, and made him accountable for all three. By the summer of 1826, at least, Karl seems to have grown more contemptuous of his uncle, and started seeing his mother clandestinely, as well as one of his 'forbidden' friends, Niemetz. It may be that this produced conflicts in him that he could not handle; there are suggestions too that he had also got into debt. On 29 July, at all events, he pawned his watch, bought two new pistols and drove to Baden. Next morning he went to the Helenenthal, one of his uncle's favourite spots, and discharged both weapons at his temple. Neither bullet penetrated the skull, and when the injured young man was found he was carried back to Vienna – to his mother's house.

Karl's attempted suicide proved shattering to Beethoven; Schindler describes him soon after as looking like a man of 70. He was urged by his friends to give up the guardianship and to reach a decision about Karl's future, for the penal aspects of the case were a constant threat. Two years earlier Karl had expressed a wish to enter the army, and now, through the help of Stephan von Breuning, it was arranged for him to be taken as a cadet into the regiment of a certain Baron von Stutterheim. Beethoven's gratitude for this outcome is shown by the fact that he changed the dedication of his C♯ minor Quartet, which he had declared to be his greatest, so that it could be dedicated to this unmusical warrior.

Since 1819 Beethoven's brother Johann had owned a country property at Gneixendorf near Krems. Beethoven had often been asked to stay, but his dislike of his sister-in-law Therese had led him to turn the invitations down; shocked by her infidelities, in fact, he had from time to time urged Johann to divorce her and to make a will leaving his fortune to Karl. On this occasion, judging it prudent to be absent from Vienna, he accepted Johann's invitation; and three days after Karl had been discharged from hospital on 25 September the two brothers travelled to Gneixendorf with their nephew, arriving after an overnight stop at a village. Beethoven was ill when he left Vienna; he seems also to have been very depressed and withdrawn, and his eccentricities of behaviour were found comic by the country folk. Yet as usual he managed to work. Since July he had been occupied with a quartet in F (op.135); he completed it at Gneixendorf by the middle of October, copied out the parts himself, and sent it straight away to Schlesinger in Paris. Then he turned to a problem that had arisen with the B♭ Quartet (op.130). Because of the difficulty that had been found with the fugue that formed its last movement, he was asked by the publisher to supply a new, easier finale (which would be paid for). After reflection he undertook to do so, and delivered it to the publisher in the middle of November. It was the last complete piece that he composed. The 'Grosse Fuge', it was agreed, should be published as well, but as a separate opus (op.133).

Beethoven started back to Vienna with Karl on 1 December, arriving there the next day, and having got to his lodgings in the building known as the Schwarzspanierhaus he immediately called a doctor. He had already had swollen feet in the country, but the underlying pathology became manifest on 13 December when he developed jaundice

7. Beethoven's study in the Schwarzspanierhaus: pen and ink drawing with wash (1827) by Johann Nepomuk Hoechle, made a few days after the composer's death

and ascites (dropsy). His doctors appear to have perceived correctly that his liver was affected (probably a post-hepatitic cirrhosis), but there was little they could do beyond relieving his swollen abdomen by tapping off the fluid. This was done on 20 December, and again on 8 January, 2 February and 27 February 1827. Meanwhile news of the seriousness of his condition, and exaggerated reports about his financial needs, had spread far and wide. The firm of Schott sent him a dozen bottles of Rhine wine; the Philharmonic Society of London resolved to provide £100 for his relief. There were occasional letters from Karl, now with his regiment, and some entertainment for the sick man was provided by Breuning's 13-year-old son Gerhard, who called daily. There was also a stream of other visitors.

When it was clear that the end was near Breuning drafted a simple will, which bequeathed Beethoven's whole estate to Karl; on 23 March Beethoven copied and signed it ('luwig van Beethoven') with great difficulty. He died at about 5.45 p.m. on 26 March. The funeral on 29 March was a public event for the Viennese; the crowd was estimated at 10,000. The funeral oration, written by Franz Grillparzer, was delivered at the graveside in the cemetery at Währing by the actor Heinrich Anschütz. In 1888 Beethoven's remains were removed, together with Schubert's, to the Zentralfriedhof (Central Cemetery) in Vienna, where they now rest side by side.

CHAPTER TWO

Works

I The 'three periods'

The division of Beethoven's life and works into three periods was adumbrated as early as 1828 by Schlosser, taken up by Fétis in 1837, and then elaborated and popularized by Lenz in his influential *Beethoven et ses trois styles* of 1852. They saw a first formative period ending around 1802, a second period lasting until around 1812 and a transcendent third period from 1813 to 1827. This schema has been attacked, not without reason, as simplistic and suspiciously consonant with evolutionary preconceptions. Yet it refuses to die, because in spite of all it obviously does accommodate the bluntest style distinctions to be observed in Beethoven's output, and also because the breaks between the periods correspond with the major turning-points in Beethoven's biography. There can be no doubt that with Beethoven – not to speak of other composers – a very close relationship existed between his creative energies and his emotional life.

The three-period framework should not be scrapped, then, but it is certainly in need of some refining. The following takes account of a number of suggestions made in the more recent literature. First, a fourth period should be added, or rather, divided off from the traditional first period: the music composed at Bonn, about which the 19th century knew little and probably cared less. Second, examination shows that each of the four periods breaks naturally into two sub-periods, and so they are best conceived of in this way. Third, allowance must be made not only for the general development of a composer's style, but also for the inner necessities of certain genres and the effects of his experience with them. For example, works in genres which he was attacking for the first time may have less 'advanced' stylistic features than works of the same date in familiar, much used genres.

It is also necessary to understand that in each of the four periods the nature of the two sub-periods and their relation to each other differ considerably. In the Bonn period the first sub-period (1782–5) contains juvenilia of small importance. Then there seems to be a pause; it is known that the years 1786–9 were very eventful ones for Beethoven but little is known of any music he composed in this period. From 1790–92 a group of much more mature works survives – a rather impressive corpus, indeed, which could reasonably support the young composer's ambitious plan of study in Vienna.

In the early Vienna period, Beethoven first had to gain control over the Viennese style and assert his individuality within it (1793–9). Then from 1800 to 1802 he produced at high speed a series of increasingly experimental pieces which must be seen in retrospect as a transition to the middle period. It is in this sub-period that the relative effects of genre and familiarity are especially clear. In 1798 and 1799 the piano sonatas are fluid and visionary but the earliest string quartets are relatively stiff. By 1800 the quartet writing moves more easily but the first of his symphonies is still decidedly conservative.

The middle period begins with a famous series of compositions in the heroic vein (1803–8): the Eroica Symphony, *Leonore* (*Fidelio*) and others. The music of the sub-period 1809–12 follows the same general stylistic impetus, but becomes rather less radical and turbulent as it becomes more and more effortless in technique. Most of Beethoven's orchestral music dates from the middle period.

The late period is in every way the most complex. In 1813–18, years marked by emotional upheavals, Beethoven's output fell off sharply. Naturally enough, most attention has been directed to the few compositions in this sub-period of a more serious nature; increasingly intimate and even 'private', they convey unmistakable hints of a new style. But the years 1813–16 also saw many 'public' works, such as the 'Battle Symphony', *Der glorreiche Augenblick* and the *Chor auf die verbündeten Fürsten* for the Congress of Vienna. These, as Maynard Solomon has observed, 'regressed to a pastiche of the heroic style' and show just as unmistakably that the style change was now being worked out slowly and with great difficulty – not at all like the earlier transition in 1800–02. The Hammerklavier Sonata of 1818 represented a kind of breakthrough, but only after the matter of his nephew's guardianship was settled by the courts did Beethoven's compositional energies flow easily again, in the unbroken series of late-period masterpieces written from 1820 to 1826.

91

II Music of the Bonn period

Ten compositions by Beethoven are known from the years 1782–5, when efforts were being made to promote him as a prodigy. Publication was gained for most of these works. Another 30 or so from the years 1787–92 are extant; of these, few appear to pre-date 1790 and only one was published at the time. As a good many of the others are known only from later sources, scholars have always suspected that they may be known in considerably revised versions. It was a pet theory of Thayer, the great 19th-century Beethoven biographer, that the young composer brought a thick portfolio of music from Bonn to Vienna and drew on it liberally for compositions of the next decade and even later. Rather more than most composers, as Thayer had observed, Beethoven was inclined to publish his juvenilia in later life and also to incorporate parts of them into mature pieces. And this in turn suggests a special motive for studying the unassuming music of Beethoven's Bonn years.

The most substantial of the earliest compositions are sets of three piano sonatas and three piano quartets. The main musical influences on the boy have been seen as, first, Neefe and Sterkel, and then Mozart; each of the piano quartets is modelled on a specific work by Mozart, from the set of violin sonatas published in 1781 (к379/373*a*, 380/374*f*, 296). Beethoven looked to Mozart again and again during his first decade in Vienna (see opp.3, 16, 18 no.5).

During the second Bonn sub-period Beethoven produced about a dozen lieder of considerable interest. He published some of them later in op.52 (1805), but only the simpler ones; the more elaborate and intense Bonn songs are not well known because they were discovered relatively late and buried in the 1888 supplement to the Gesamtausgabe. In 1790, the important commission to prepare official cantatas on the death of Emperor Joseph II and the accession of Leopold II spurred Beethoven on to the most ambitious of his youthful projects. The funeral cantata gave him the opportunity for some admirably expressive writing in the pathetic C minor chorus which frames the work and in the serene soprano aria with chorus. He was to use this again with superb effect at the dénouement of *Leonore*, 15 years later. In addition to the five large arias within these cantatas, he also composed three accomplished concert arias: *Prüfung des Küssens*, *Mit Mädeln sich vertragen* (his first Goethe setting) and *Primo amore*.

A genre in which any budding virtuoso had to excel was the variation set. In 1790–92 Beethoven wrote out two brilliant sets for piano, on Righini's 'Venni amore' and Dittersdorf's 'Es war einmal ein alter Mann'; one set for piano duet on a theme by Count Waldstein; and one for violin and piano on Mozart's 'Se vuol ballare' (completed in Vienna). While many of the variations are of the insipid decorative variety, others deal with the theme in a more interesting, substantive fashion. It is in these 'analytical' variations, perhaps, more than in the other Bonn music, that the Beethoven to come can be glimpsed.

Less impressive, in these years, is the instrumental music in the sonata style. There is an incomplete draft for a passionate symphony movement in C minor;

fragments of a big violin concerto and of some sort of concertante for piano, flute and bassoon; a complete trio for the same three instruments; a piano trio (WoO 38) and what looks like part of a movement from another, and a few rather colourless sonata movements for piano. There are also many sketches. (Is it accidental that so much of this music has been transmitted in an incomplete form? An oboe concerto and the original version of the B♭ Piano Concerto, both dating from this period, have vanished with barely a trace.) Where Beethoven departed from formula in these works he seems to have straggled helplessly, as in the violin concerto fragment. Although there are some bold strokes, they are seldom integrated convincingly into the total musical discourse.

Greater sophistication is shown by the Wind Octet op.103, but here there is reason to believe that Beethoven rewrote what was originally a Bonn score during his first years in Vienna, with an eye to publication. Leaving this work out of consideration, one is bound to conclude that Beethoven at Bonn was a less interesting composer of works in the sonata style than of music in other genres – variations, lieder and large vocal-orchestral pieces. In view of his later output, this conclusion may seem surprising. Yet the sonata style as it is generally known was very much a Viennese speciality. The Bonn works in the sonata style make clear how important and right it was for Beethoven to have gone back to Vienna in late 1792, and how large a part Vienna was to play in the formation and nurture of his musical personality.

III Music of the early Vienna period

During his first year or so in Vienna Beethoven appears to have composed considerably less than in the years just preceding and following. There are signs that he spent some time revising or recasting an amount of his Bonn music to reflect Viennese standards and taste. The Wind Octet has already been mentioned; sketches show that he also started reworking his violin and oboe concertos. Fragments of the juvenile piano quartets were incorporated into some of the first sonatas composed in Vienna, op.2 nos.1 and 3.

By the time opp.1 and 2 were published (July 1795, March 1796) Beethoven certainly had the musical wherewithal to make Vienna sit up and listen. Probably the best-known movement from this impressive group of six pieces is the opening Allegro of the Piano Sonata in F minor op.2 no.1, a remarkable precursor of Beethovenian concentration and intensity (and the more remarkable in that the sketches go back to Bonn). In 1795, however, this movement was an exception. Most of the early music is scaled very broadly, weighty and discursive, even overblown. Thus for many years Beethoven almost invariably wrote sonatas in four movements, rather than three, as was the rule with Haydn and Mozart, and it seems indicative that his op.3 was a string trio in six movements, modelled on the large Divertimento K563 by Mozart. There is inconclusive evidence that op.3 goes back to a Bonn original, but in its final form it was certainly written in Vienna, like the Wind Octet.

Opp.1 and 2 provide examples of the rather ponderous slow movements characteristic of the first Vienna period, and also of that famous innovation the scherzo. Beethoven's early scherzos move no faster than most Haydn minuets and sound no more humorous, but they last considerably longer and tend to be constructed out of more symmetrical periods. As for movements in sonata form, most of them contain a great deal of musical material – and a great many modulations in the second group. Though Beethoven's still emerging powers of organization were sometimes overtaxed, sometimes they were not and there are passages of authentic Beethovenian power, especially in the matter of long-range control over bold harmonic action. Cases in point are the passing modulations in the first movement of the A major Sonata op.2 no.2, and the expanded recapitulation in the Adagio of the G major Trio op.1 no.2.

In these early years Beethoven made his name as a pianist and improviser and as a composer primarily for piano. Some ideas of his improvising style can be formed from his published piano variations, from copious notations on his early sketchleaves, and from certain incomplete piano scores which are perhaps better viewed as *aides-mémoires* than as unachieved compositions. The well-known *Rondo a capriccio* was completed by Diabelli after Beethoven's death and published as op.129 under the irresponsible title 'Rage over a Lost Penny', and an interesting cyclic 'Fantasia' in three movements has recently come to light in the so-called 'Kafka sketchbook' (British Library). In later years he improvised less, of course, but evidence of his style is still to be found in the Fantasias opp.77 and 80, the cadenzas to piano concertos, and shorter cadenza-like passages in a very large number of other pieces.

Beethoven was naturally open to the influence of other pianist-composers at a time when the technique of the instrument was expanding significantly. Too much can be made, however, of similar themes and pianistic textures in Beethoven and Clementi, Dussek and other such composers. From the start, and even at his most discursive, Beethoven had a commitment to the total structure that makes Clementi seem very lax. His well-known insistence on making transitional and cadential matter sound individual is already in evidence; he had little use for the debased coin of the *style galant* which was still in circulation in the 1790s. And in his 'serious' compositions piano virtuosity is always used in the service of a musical idea, never for its own sake. These compositions may sound pompous or gauche, sometimes, but they never sound meretricious and they never lack a certain intellectual and imaginative quality.

As has been mentioned above, when Haydn heard the op.1 trios he praised them but thought the public would not understand or accept the third, in C minor. One suspects that Haydn himself may have been put off by the extremes of tempo, dynamics, texture and local chromatic action in this piece, and still more by the resulting emotional aura. He would not have been the last listener to find something callow and stagey, which is to say essentially impersonal, in these insistent gestures of pathos and high drama. Beethoven of course paid no attention to his advice and published increasingly sophisticated C minor items in nearly every one of his composite sets of works over the next eight years (opp.9, 10, 18, 30). In these years C minor was practically the only minor key he used for full-length pieces (though D minor is used for the impressive slow movements of op.10 no.3 and op.18 no.1, as well as for the 'Tempest' Sonata op.31 no.2). The most successful early embodiments of Beethoven's 'C minor mood' are no doubt the *Sonate pathétique* op.13 (1799) and the Third Piano Concerto (?1800). Still to come were the 32 Variations on an Original Theme, for piano, the *Coriolan* Overture, the Fifth Symphony and the last piano sonata.

The first movements, in sonata form, of the C minor Trio and the F minor Sonata have quiet main themes which are designed to return *fortissimo* at the point of recapitulation. This is a characteristic Beethoven fingerprint. In the early works it often makes for a rather blustery effect. Yet it adumbrates a new view of the form whereby the recapitulation is conceived less as a symmetrical return or a climax than as a transformation or triumph. The sonata style is always inherently 'dramatic', in the special sense expounded and illuminated by Tovey. Tovey also pointed out that at their most characteristic Haydn and Mozart use the style to project high comedy, the musical equivalent of a comedy of manners. Beethoven was already groping for ways of using it for tragedy, melodrama or his own special brand of inspirational theatre of ideas.

This radical approach to sonata form (which encompasses all its aspects, of course, not only the enhanced recapitulation) becomes clearer in the piano sonatas of

1796–9: op.7, op.10 nos.1–3 and op.13. In op.13 and in the fine Sonata in D op.10 no.3, although the main theme does not return loudly, there is still a compelling impression that something urgent is at stake in the musical dialectic. Broadly speaking, it was this sense of urgency in dealing with the Classical style that Viennese aristocratic circles found most novel and impressive in the 'grand Mogul', as Haydn called him, from the provinces.

A deliberate campaign to annex all current musical genres can be read into Beethoven's activities in these years. He wrote an effective concert aria – a scena and rondò – to a text adapted from Metastasio, *Ah! perfido*, some deft little songs to lyrics by Goethe, and an interesting extended lied, the once-popular *Adelaide*. He produced two rather Mozartian piano concertos, one of them (the B♭, op.19) evidently revised several times from a Bonn original, and a good deal of miscellaneous wind music, including a Quintet for Piano and Wind op.16 which incautiously invites comparison with a similar work by Mozart (κ452). In 1795–6 he sketched long and hard at a symphony in C. As it was turning out to be too big, he wisely shelved it, though he returned to some of its musical ideas when he wrote the First Symphony (also in C) in 1800.

The three Violin Sonatas op.12 are not as impressive as the contemporary piano sonatas; the two Cello Sonatas op.5 are also lesser works but interesting in their bold virtuoso stance, looking ahead to the Kreutzer Sonata of 1802–3. After completing the three String Trios op.9 Beethoven launched into his most ambitious project yet, the set of six String Quartets op.18 (1798–1800). All the while he was contributing copiously to the ephemera of Viennese musical life: easy piano variations, ballroom dances by the dozen, patriotic marching songs, arias to be inserted into a Singspiel, pieces for mechanical clock-organ and a Sonatina for mandolin and piano.

There is no single work that demarcates the second sub-period within the early Vienna years, the time when Beethoven began to show signs of dissatisfaction with some of the more formal aspects of the Classical style and reached towards something new. In a way the signs were present from the beginning. Novelties of conception can be detected all along. They are multiplied in the *Sonata pathétique* of 1799 – the integration of the introduction into the first movement proper, the perfectly managed bold modulations in the second group, the prophetic breakdown on the dominant in the middle of the rondo; not to speak of the overall coherence of mood which has made the *Pathétique* the most famous piece in Beethoven's early output. Another famous early piece, the first movement of the Quartet in F op.18 no.1, is Beethoven's first exhaustive study in motivic saturation. The turn-motif of bars 1–2 forces its way into every available nook and cranny of the second group, the transitions, the development and coda. (When Beethoven revised op.18 no.1, after having given a fair copy of it to Amenda, he reduced the appearances of the turn-motif by nearly a quarter.)

The last two quartets of op.18, composed around 1800, show a rather new treatment of the traditional four-movement form. The first movements are not extensive and decisive but instead swift, bland and

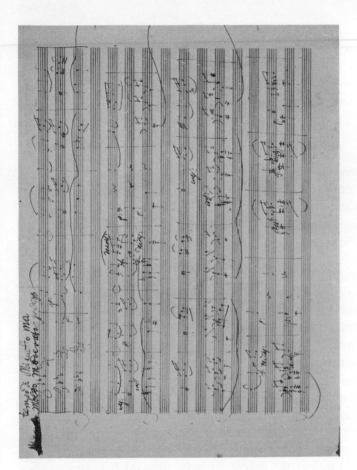

8. Autograph MS of
the opening of the
2nd movement of
Violin Sonata
in G op.30 no.3,
composed in 1802

symmetrical, so that the later movements all seem (and were surely meant to seem) weightier or more arresting. The most visionary of these later movements is the composite finale of the Quartet in B♭ op.18 no.6, where a slow, strange-sounding chromatic labyrinth entitled 'La malinconia' alternates with a swift, limpid little dance evocative of the Viennese ballrooms.

More far-reaching experiments with the weight, character and balance of the various movements in a work were made within the impressive series of about a dozen sonatas composed in 1800–02. These included op.26 in A♭ with its 'Marcia funebre sulla morte d'un Eroe' (cf the Eroica Symphony), op.27 no.1 in E♭, op.27 no.2 in C♯ minor (each marked 'quasi una fantasia') and op.31 no.3 in E♭. Some of the movements are run together, and there is a significant shift in weight away from the first movement and towards the last. Experiments of this kind with multi-movement works slowed down during the next period, when under the sway of his developing 'symphonic' ideal Beethoven found fresh resources in the traditional four-movement dynamic. But they played an important part in the growing flexibility of his art, and after 1812 they were resumed with much greater force and consciousness.

Greater flexibility already allowed for the incorporation of movements of widely different characters and forms. It is perhaps at this time that one first begins to be aware of the striking individuality of all Beethoven's pieces, a characteristic that has often been noted. Prime examples are the so-called 'Pastoral' Sonata in D op.28 (also the *locus classicus* for successive thematic fragmentation in a development section), the Sonata in E♭ op.31 no.3, and those great and deserving favourites of the Romantic era, the *Pathétique*, the 'Moonlight' and the D minor op.31 no.2.

The opening reverie of the 'Moonlight' is such a startling conception, even today, that Beethoven's very careful plotting of the sequence of the movements in this sonata seems to pale by comparison. Unprecedented for a sonata opening is the half-improvisatory texture, the unity of mood, and especially the mood itself – that romantic *mestizia* which will have overwhelmed all but the stoniest of listeners by the end of the melody's first phrase. An equally bold and emotional, but also more intellectual, experiment marks the opening of op.31 no.2. Here the first theme in a sonata form movement consists of antecedent and consequent phrases of radically different characters: a slow improvisatory arpeggio and a fast, highly motivic *agitato*. Both of these ideas can be heard echoing in the later movements of the sonata.

The inner pressure of his developing musical thought drove Beethoven on to more and more novelty, no doubt; and mixed in with this was a measure of artistic vanity. About 1801–2 he appeared much concerned with being original, even advising a publisher to point out the innovations in his Piano Variations on Original Themes opp.34 and 35 by means of a special advertisement. And that would certainly have been justified. Op.34 has its six variations in six different keys. Op.35 abstracts the ludicrous bare bass line of the contredanse theme from *Prometheus* and builds up from it fantastically in 15 variations and a full-length fugue.

The finale of the Eroica is a second building exercise on the same bass, this time involving variations in different keys and two fugato sections.

According to Czerny, his young pupil in those years, Beethoven spoke of a 'new path' he was following, a path which later Czerny associated with the important op.31 sonatas of 1802. Mention has already been made of op.31 no.2. Another significant novelty of conception was the key plan of the first movement of the Sonata in G op.31 no.1, which has the second group not in the dominant but in the mediant key (major and minor; cf the String Quintet op.29 of 1801). This looks ahead to Beethoven's thorough exploration and extension of the tonal range of Classical music, a process that was to run parallel with his expansion of all aspects of Classical form in the next years. In the late period it is the exception rather than the rule to have the second group in the dominant.

Other important, but more conservative, works of 1799–1801 are the music for the ballet *Die Geschöpfe des Prometheus*, Beethoven's introduction to the Viennese stage; the amiable but rather mindless Septet, whose great popularity soon came to irritate the composer; and the slender First Symphony, which can seem almost to wilt when commentators examine it for clues to future symphonic greatness. The Second Symphony of 1802 must also be counted among the more conservative works – this in spite of its great advance in assurance over the First and its inspired play with the notes F♯ and G as a means of unifying the whole. Although Haydn could never have written this work, it stands as a final ideal realization of the concept of a large concert piece which he had developed. This impression is confirmed by Beethoven's quotation of a sensational modulatory passage from *The Creation*, as Tovey observed.

One feels that in the Second Symphony Beethoven for the first time really engaged with the symphony orchestra and began to understand how it could serve his own emerging purpose. He had taken its true measure. In the middle period, from 1803 to 1812, he wrote most of his famous works for orchestra, evolving through them a new 'symphonic ideal' that also inspired most of his non-orchestral music.

IV The symphonic ideal

After the period of inner turmoil expressed (and perhaps resolved) by the Heiligenstadt Testament of October 1802, Beethoven began to engage seriously with large works involving extra-musical ideas. It was the first time he had done so since going to Vienna. An outer impetus was his association first with the Burgtheater and then with the Theater an der Wien, but the decision to embark on a 'Bonaparte symphony' at just this time came from inner pressures. The oratorio *Christus am Oelberge*, musically not a great success, was written hastily in early 1803. The opera *Leonore* was written very slowly in 1804–5. Between them came the Eroica Symphony, no.3: an authentic 'watershed work', one that marks a turning-point in the history of modern music.

Thanks to Nottebohm's monograph on the Eroica sketches, more is generally known about the composition of this work than any other by Beethoven. The sketches show a minimum of false starts and detours. The most radical ideas were present from the start, if in cruder form, and work seems to have proceeded with great assurance. This is striking indeed, for however carefully one studies Beethoven's evolving style up to 1803, nothing prepares one for the scope, the almost bewildering originality and almost continuous technical certainty manifested in this symphony. In sheer length, Beethoven may well have felt that he had overextended himself, for it was many years before he wrote another instrumental work of like dimensions.

In the first movement, one must marvel at the expansion in dimensions on every level; at the projection of certain melodic details of the main theme into the total form – the bass $C\sharp$ ($D\flat$) instigating moves to the keys of the supertonic and the flat seventh degree in the recapitulation, the violins' G–$A\flat$ returning vertically as the famous horn-call dissonance; at the masterly coagulation of diverse material into the second group; and at the whole concept of the panoramic development section, with its passage of deepening breakdown redeemed by the introduction of a new theme (if it is indeed really new). The moving thematic 'liquidation' at the end of the Marcia funebre, the four *alla breve* bars in the da capo of the scherzo, the novel structure of the finale, the powerful fugatos throughout – none of these could have been predicted. Also astonishing is the quality of 'potential' that informs the main themes of the three fast movements. Two of them require (and in due course receive) horizontal or vertical completion, and the other is presented in a state of almost palpable evolution.

These themes were made to order for the new 'symphonic ideal' which Beethoven perfected at a stroke with his Third Symphony and further celebrated with his Fifth, Sixth, Seventh and Ninth. The forcefulness, expanded range and evident radical intent of these works sets them apart from symphonies in the 18th-century tradition, such as Beethoven's own First and Second. But more than this, they all contrive to create the impression of a psychological journey or a growth process. In the course of this, something seems to arrive or triumph or transcend – even if, as in the Pastoral, what is mainly transcended is the weather. This illusion is helped by certain other characteristic features: 'evolving' themes, transitions between widely separated passages, actual thematic recurrences from one movement to another, and last but not least, the involvement of extra-musical ideas by means of a literary text, a programme, or (as in the Eroica) just a few tantalizing titles.

In technical terms, this development may be viewed as the projection of the underlying principles of the sonata style on the scale of the total four-movement work, rather than that of the single movement in sonata form. This view takes account of the impression Beethoven now so often gives of grappling with musical fundamentals. He had the power – and it must be called an intellectual power – of penetration into the gestural level below sonata form. He could manipulate the basic elements of the sonata style in a more comprehensive, less formalistic way than ever before. One senses the same grasp of essences when Beethoven now isolates a melodic, harmonic or rhythmic detail of a theme and then appears to 'compose it out' – to spell out its implications later in the piece. Doubtless this also happens in earlier music, by Beethoven or by other composers, but in the middle period he began to draw attention to the process in a much more pointed fashion.

Beethoven's fascination for musicians of a certain turn of mind rests on his continuing investigation of basic musical relationships in this sense. The investigations grew more momentous in the late period, and also more subtle and pervasive, as will be seen if one compares the 'composing out' of C♯ and A♭ in the Eroica first movement, mentioned above, with the treatment of the Neapolitan D in the Quartet in C♯ minor op.131.

For musicians and listeners of another turn of mind, Beethoven's attraction rests on another aspect of the 'symphonic ideal', one that is less technical but probably no less essential. The combination of his musical dynamic, now extremely powerful, and extra-musical suggestions invests his pieces with an unmistakable ethical aura. Even Tovey, the most zealous adherent of the 'pure music' position, was convinced that Beethoven's music was 'edifying'. J. W. N. Sullivan taught the readers of his influential little book to share in Beethoven's 'spiritual development'. Concert-goers of the 19th and 20th centuries gladly attached programmatic suggestions to those symphonies that lack them: to the Fifth, Beethoven's remark about fate knocking at the door, and to the Seventh, Wagner's less happy evocation of an apotheosis of dance. In 1937 the eccentric musicologist Arnold Schering proposed detailed Shakespearean and other literary programmes for a whole clutch of Beethoven compositions.

An important influence on the Eroica Symphony and other works of this period is that of French post-revolutionary music. In 1802 and 1803 operas by Cherubini and Méhul played in Vienna for the first time, with enormous success. Their impact on Beethoven has been traced in such diverse areas as his driving orchestral tutti style, his partiality for marches and march-like material, the free form of his overtures (*Leonore* no.2, 1805, stands in the same relation to *Prometheus*, 1801, as the Eroica Symphony does to the First), and various points of harmony and orchestration. Beethoven's symphonic ideal itself is foreshadowed in the French repertory of the 1790s, in the grand revolutionary symphonies, sometimes with chorus, by Gossec, Méhul and their contemporaries. But with Beethoven there is not only an incomparably more arresting musical technique but also a decisive change in emphasis. He personalized the political symphony. The Eroica was conceived as a tribute not to the idea of revolution but to the revolutionary hero, Napoleon, and really to Beethoven himself. Later concert-goers have been able to respond to Beethoven's spiritual journeys in a way they could never respond to celebrations of long-past political ideologies.

The conception of this symphonic ideal, and the development of technical means to implement it, is probably Beethoven's greatest single achievement. It is *par excellence* a Romantic phenomenon, however 'Classical' one may wish to regard his purely musical procedures. It is also a feature that has offended certain critics, especially in the early part of this century, and set them against Beethoven. The composer himself was capable of producing a cynical and enormously successful popular travesty of his own symphonic ideal, in the 'Battle Symphony' of 1813.

V Middle-period works

Soon after the Eroica Symphony the Fifth was conceived, but somehow work got deflected into certain other C major and minor projects, and things did not come together until late 1807 and 1808. More than any other piece of music, the Fifth Symphony has come to typify the thematic unification, or 'organicism', as the 19th century viewed it, that Beethoven developed to such a high degree in these years. The famous opening motif is to be heard in almost every bar of the first movement – and, allowing for modifications, in the other movements. The opening theme expands into the horn-call before the second subject, and the second subject employs the same note pattern as the horn-call. Then, in the development section, the horn-call is fragmented successively down to a single minim, alternating between strings and woodwind in a passage of extraordinary tension achieved primarily by harmonic means. As in most other works of the time, the last two movements are run together without a break; this device, obviously, contributes to the continuity and to a feeling of necessary sequence. But more than this: here the long transition passage between the movements, and the recurrence of a theme from the third movement in the retransition before the recapitulation of the fourth, give the sense that one movement is triumphantly resolved by the other – a sense confirmed by the enormously emphatic last-movement coda.

Such codas now become very common. They tend to assume the important function of finally resolving some melodic, harmonic or rhythmic instability in the first theme – an instability that has infused the movement with much of its energy up to the coda. This new weighting of sonata form towards the coda is associated, and sometimes coordinated, with another tendency, that of withholding full rhythmic or even harmonic resolution at the moment of recapitulation. Thus in the first movement of the Fourth Symphony, as Robert Simpson has observed, solid dominant–tonic resolution waits in the recapitulation until the appearance of the second theme (compare the first movements of two other works in the same key, B♭, the Hammerklavier Sonata and the Quartet op.130). The Fourth Symphony, said Tovey, 'is perhaps the work in which Beethoven first fully reveals his mastery of movement'.

Hardly less original than the Fifth Symphony is the Sixth (Pastoral, 1808), though here for once the first movement is made as quiet as possible. This is done with the help of a development section devoid of tensions, a recapitulation approached hymn-like from the subdominant, and countless pedal points throughout. In compensation, a passage of fury comes elsewhere in the piece, as an extra movement (trombones and piccolo enter for the first time in the symphony to enforce this 'Storm'). Each of the five

111

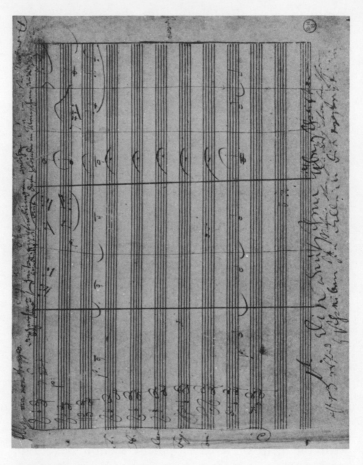

9. Autograph MS
of the opening of
Symphony no.6
(Pastoral),
composed 1807–8

112

movements bears a programmatic inscription, and one of these is frankly pictorial in nature – the 'Scene by the brook' inscribed over the slow movement, which includes a series of stylized birdcalls at the end, in a sort of woodwind cadenza (Beethoven was careful to identify the quail, nightingale and cuckoo). On the other hand, he stressed the word 'Gefühle' ('feeling') in two other inscriptions and so could quite properly observe that his reference was less to musical 'Malerei' ('painting') than to emotions aroused by the countryside. A sequence of such feelings guides the listener through the familiar therapeutic progress of a Beethoven symphony, in a somewhat gentler version.

The symphonic ideal inspires most of the nonsymphonic pieces written between 1803 and 1808. That is true to an extent even of the Kreutzer Sonata, composed in early 1803, just before the Eroica. The Waldstein Sonata, composed just after the Eroica, adopts an idea for the groundplan of its opening paragraph from an earlier piano sonata, op.31 no.1 in G. But there is all the difference in ambition, scale and mood; what served in the earlier piece as a witty constructive device becomes in the later one an earthshaking, or at least a piano-shaking, declaration. The slow movement was originally going to be the somewhat bovine piece now known as the 'Andante favori' (compare the Kreutzer and op.31 no.1). When Beethoven replaced this by the *adagio* 'Introduzione' which makes momentous preparations for the finale, he gave the sonata the characteristic 'symphonic' sweep even while shortening it, and also motivated (or validated) the grandiose coda of the finale. Planned on broader lines still, the 'Appassionata' Sonata (1804–5) is an

113

even more imaginative work, a work of the greatest extremes – as witness the *fortissimo* chord handfuls that shatter the brooding quiet of the very first page.

This and other equally violent effects were hardly thinkable on the Walter fortepiano owned by Beethoven before 1803, when he got his Erard (now in the Vienna Kunsthistorisches Museum). Yet even when dealing with instruments that were not in a state of radical development, he acted as if they were. The string quartets of op.59 so strained the medium, as it was understood in 1806, that they met with resistance from players and audiences alike. The first movement of the F major Quartet op.59 no.1, though in mood very different from the Eroica Symphony, resembles it in its unexampled scope and also, rather surprisingly, in a number of technical features. The second movement is Beethoven's largest, most fantastic scherzando – not a true scherzo, but a free essay in the tradition of the sonatas op.31 no.3 and op.54. All three quartet slow movements, surely, cry out for evocative titles, and the last two finales are all but orchestral in conception.

Each quartet was supposed to include a Russian melody, for the benefit of the dedicatee Count Razumovsky, the Russian ambassador to Vienna. Here for the first time may be seen Beethoven's interest in folksong, which was to grow in later years. Folksongs did not much help the first two quartets, but Razumovsky's notion came to superb fruition in the third, where Beethoven gave up the idea of incorporating pre-existing tunes and instead wrote the haunting A minor Andante in what he must have conceived to be a Russian idiom.

In some ways the 1805 *Leonore* stands apart from other major works of these years. In local musical terms, the innovations and expanded horizons of the instrumental works are not deeply reflected in the separate operatic numbers, and probably could not have been. Apart from the overtures, there is a certain stiffness about many numbers which is understandable in a first opera. This quality is also discernible in Beethoven's first oratorio and mass, *Christus am Oelberge* (1803–4) and the Mass in C (1807).

In broader musical terms, however, the importance of *Leonore* can scarcely be exaggerated. Faced by the task of matching music to an explicit narrative, and doubtless instructed by the Mozart operas which we know he consulted at the time, Beethoven here estab-lished a very large-scale dramatic continuity largely by tonal means. The *Leonore* overtures are famous for forecasting (and, in a sense, giving away) the opera's turning-point by incorporating the trumpet signals for the arrival of the Minister who confounds the villain Pizarro. But the overtures also assert C major as the opera's tonic key and A♭ and E as subsidiary keys, and *Leonore* no.3 precedes its final triumphant tonic sec-tion with a recapitulation in G major; then the C minor/major of the first vocal number (see Table 1) leads through many detours to moments central to the drama in E and A♭ and then to G major, C minor and C major in the last two numbers. Even more than the Eroica Symphony, *Leonore* prefigures the more ab-stract (and of course more concise) tonal structures of the later instrumental works.

In terms of idea, furthermore, *Leonore* provides a

115

shining prototype for the heroic progress implied in a less explicit way by the instrumental music. And what is remarkable is to see Beethoven gradually evolving a personal operatic style in the course of writing, and rewriting, *Leonore*. From the somewhat servile echoes of French and German light opera in the opening numbers, he moved on to find an increasingly individual and elevated voice – for example, in the Prisoners' Chorus, the scena for Florestan, the duet 'O namenlose Freude' (revised from the *Vestas Feuer* fragments of 1803) and the long recitative before it which was the most regrettable of Beethoven's cuts for the 1814 version (for a comparison of the versions see Table 1). To say that Beethoven approached his libretto with utter seriousness and idealism may seem like a truism; but of how many other first operas of the time can as much be said?

Around 1808 the enthusiasm and high daring of Beethoven's music begins to be tempered by ever-increasing technical virtuosity. Even when the pieces are still very powerful, as is often the case, they are smoother and a little safer than before. The stage work of this period is *Egmont* (1809–10), consisting not only of the well-known overture but also incidental music lasting 40 minutes, including a final 'Siegessymphonie' ('Symphony of Victory') in the face of disaster. Feelings that were turned inward in *Leonore* were turned outward in *Egmont*. Whereas the *Leonore* no.2 and no.3 overtures were involuted, explosive works dedicated to gigantic struggle, the *Egmont* overture is a tough, lucid one that comes by its Pyrrhic victories easily.

116

TABLE 1: The three versions of Fidelio (Leonore)

1805	1806	1814
Leonore (Joseph Sonnleithner, after J. N. Bouilly: Léonore ou L'amour conjugal)	Leonore (1805, rev. Stephan von Breuning)	Fidelio (1806, rev. Georg Friedrich Treitschke)
Overture: Leonore no.2, C	Overture: Leonore no.3, C	Overture: Fidelio, E
Act 1	**Act 1**	**Act 1**
1 Aria (Marzelline) 'O wär ich schon', c/C	1 1805/1	1 1805/2
2 Duet (Marzelline, Jaquino) 'Jetzt, Schätzchen', A	2 1805/2	2 1805/1
3 Trio (Marzelline, Jaquino, Rocco) 'Ein Mann ist bald genommen', Eb	3 1805/4	3 1805/4
4 Quartet (Marzelline, Leonore, Jaquino, Rocco) 'Mir ist so wunderbar', G	4 1805/6	4 1805/5
5 Aria (Rocco) 'Hat man nicht auch Gold beineben' Bb	5 1805/7	5 1805/6
6 Trio (Marzelline, Leonore, Rocco) 'Gut, Söhnchen, gut', F	6 1805/8	6 1805/7
Act 2	7 1805/9	7 1805/8
7 March, Bb (possibly not included)	8 1805/11	8 1805/9
8 Aria (Pizarro) with chorus 'Ha! welch ein Augenblick', d/D	9 1805/10	9 1805/11 with new recit 'Abscheulicher, wo eilst du hin?'
9 Duet (Pizarro, Rocco) 'Jetzt, Alter', A	10 1805/3	
10 Duet (Marzelline, Leonore) 'Um in der Ehe froh zu leben', C	11 1805/12	10 new finale
11 Recit and Aria (Leonore) 'Ach brich noch nicht ... Komm, Hoffnung', 7E		**Act 2**
12 Finale (Marzelline, Leonore, Rocco, Jaquino, Pizarro, Prisoners), Bb	**Act 2**	11 1805/13 with new final section to aria 'Und spür ich nicht linde', F
Act 3	12 1805/13	
13 Recit and Aria (Florestan) 'Gott! welch Dunkel hier ... In des Lebens Frühlingstagen', f/Ab/f	13 1805/14	12 1805/14
14 Melodrama and Duet (Leonore, Rocco) 'Nur hurtig fort', a	14 1805/15	13 1805/15
15 Trio (Leonore, Rocco, Florestan) 'Euch werde Lohn', A	15 1805/16	14 1805/16
16 Quartet (Leonore, Florestan, Rocco, Pizarro) 'Er sterbe!', D	16 1805/17	15 1805/17 without recit
17 Recit and Duet (Leonore, Florestan) 'Ich kann mich noch nicht fassen ... O namenlose Freude', G	17 1805/18	16 1805/18 but largely rewritten
18 Finale (Leonore, Marzelline, Rocco, Florestan, Jaquini, Pizarro, Fernando, Prisoners, People), c–C		

Note: The evidence of the published librettos and sketches shows that 1805/11, 13 and 14 may have been substantially changed in 1806; other movements were revised in 1806 and 1814.

117

The change is clearest of all between the op.59 quartets and the 'Harp' Quartet of 1809 (a nickname deriving from its insistent functional pizzicatos). Nothing about this work is problematic. The climax of the first movement is a climax of sheer technical exhilaration, for in the coda Beethoven seems at last to have solved the problem of simulating orchestral idiom in a quartet. The second movement is serene and the third (in C minor) sounds like a speeded-up but smootheddown version of the third movement of the Fifth Symphony. The finale is a set of simple variations on a suave 2/4 tune. This type of light finale recurs in the Violin Sonata op.96 (1812).

There are now no 'symphonic' sonatas, except perhaps the small-scaled 'Lebewohl' op.81*a* (1809–10). Beethoven's new concern in the first movements of sonatas and chamber music is lyricism, which inspires works of such different character as the Piano Sonata in F♯ op.78 (1809), the 'Archduke' Trio (1810–11) and the Violin Sonata op.96. Beethoven had never written such beautiful slow movements as he now wrote for the 'Harp' Quartet, the 'Archduke' Trio and the Fifth Piano Concerto (1809). The so-called 'Emperor' is by far the most 'symphonic' of his concertos and one of the strongest works he conceived. Yet in the very first bars, where the soloist and tutti join in a thunderous cadential celebration, the battle seems to be won even before the forces have been drawn up – as was certainly not the case in the introverted, searching Fourth Concerto first performed in 1807.

Writing his Seventh Symphony in 1811–12, Beethoven again reached for new horizons: the expanding introduction, the 6-4 chords spanning the Allegretto, the rolling ostinatos at the ends of the outer movements, the rhythmic preoccupation throughout. (For the Allegretto, he reached back to a theme jotted down in 1806 in connection with the Quartet op.59 no.3.) The Seventh Symphony is perhaps less immediate in its emotional effect than the Eroica or the Fifth, but its élan and its effortless control over musical processes at every level can make those earlier works seem more than a little hectic. The finale, all sinew, represents a particular advance, not only in elegance but also in sheer power.

Beethoven immediately capped this work with the delightful Eighth Symphony (1812), a salute to the symphonic ideal of the previous age. It has a comical slow movement and a slowish minuet in place of the now customary scherzo. Flashes of middle-period power occur only in the outer movements. Beethoven could hardly have planned a more genial gesture of farewell for a time to the symphony and to the decade of work produced under its aegis.

Another of the greatest works written between 1808 and 1812 refuses to fit any norms one may try to adduce for this period or, indeed, for any other – the Quartet in F minor op.95. (It was not published until 1816; was it possibly retouched or even in part recomposed later than 1810, the date on the autograph?) The piece is unmatched in Beethoven's output for compression, exaggerated articulation, and a corresponding sense of extreme tension. The harmonic layout is radical. Like op.57 and op.59 no.2, the first movement treats Neapolitan relationships, both in the first group

(F–G♭) and in the second (D♭–E♭♭ or D♮). The D is the key of the second movement, one of Beethoven's most beautiful, as well as one of his most disturbed – D major shadowed by D minor, with a chromatic fugato plunging into enharmonic mysteries. The F minor scherzo has a trio ranging from G♭ to D and B minor.

This *quartetto serioso*, as Beethoven called it, looks back to the impressive minor-mode compositions of the period 1803–8 and looks forward to the style and mood of the late quartets. It was some time, however, before this promise of a new style could be realized.

VI Late-period style

For a considerable time after 1812, Beethoven's production of important works fell off strikingly. These were difficult years for him, encompassing deep emotional turmoil and endless lesser distractions. In addition, he was probably suffering from something like exhaustion after the truly immense labours of the previous period. To speak only of the decade from the Eroica Symphony to the Eighth, he had composed some 30 major works which in most cases involved serious rethinking of musical essentials. He had composed nearly as many slighter works and he had seen about 80 items through the press. Long or short, great or slight, they all required negotiations with publishers, correction of copyists' scores, and proofreading – unfortunately an activity that Beethoven never fully mastered.

But more generally, these were difficult years for any serious composer of Beethoven's generation. One can perhaps appreciate the growing sense of uncertainty that he must have felt as to artistic ends and means. On some level he was responding to powerful musical currents, which were soon to come flooding to the surface; the last works of Weber and Schubert and the first works of Berlioz, Chopin and Bellini all appeared during the 1820s. Like other great composers whose lives bridged a time of deep stylistic change – like Josquin, Monteverdi and Schoenberg – Beethoven was facing a major intellectual challenge, whether or not he formulated it in intellectual terms. He had already met one such challenge, or one part of the challenge, by his reinterpretation of the sonata principle in his 'symphonic' works of 1803–12. Now the very basis of the sonata style was thrown in doubt. Beethoven had no easy answer. There is something private and problematic about the corpus of late-period works, and it is hardly accidental that their deep influence on the course of music came only much later, past the time of Beethoven's own younger contemporaries who learnt so much from the middle period.

Beethoven's concern for lyricism deepened throughout the late period. He has sometimes been criticized as an inept melodist, and it will be granted that when he was 23 he could not, like Rossini at a like age, produce the deathless melodies of a *Barbiere*. Yet some of his early Bonn songs make impressive lyric statements, and in the mid-1800s he developed a very effective type of slow hymn-like melody. This is continued, intensified and much refined in the late period; the melodic outline of Leonore's 'Komm, Hoffnung' (1805) recurs in the Adagio of the Quartet op.127 (1825). A new feature is the intimacy and delicacy already apparent in the Violin Sonata in G op.96 (1812), the Piano Sonatas in E minor op.90 (1814) and A op.101 (1816) and the Cello Sonata in C op.102 no.1 (1815).

There is also a growing interest in folklike melody, hardly surprising in one who made arrangements of over 150 folksongs for Thomson in these years. The song cycle *An die ferne Geliebte* op.98 (1816) marks Beethoven's closest approach to Goethe's ideal of the *Volksweise* as a basis for song composition (closest except for the tiny *Ruf vom Berge* WoO 147, 1816, which adapts an actual folksong melody). Simple little tunes evocative of folksong and folkdance are constantly turning up in the late quartets and other music.

In all this Beethoven appears to have been reaching for a more direct and intimate mode of communication. Two verbal adjuncts to such folklike essays can be regarded as symbolic: in the song cycle, the line 'ohne Kunstgepräng' erklungen' ('sounding without the adornments of Art'), and in the Ninth Symphony, Schiller's famous apostrophe to universal brotherhood. In the best early Romantic spirit, Beethoven was seeking a new basic level of human contact through basic song, as though without sophistication or artifice. Another manifestation of this powerful – and sometimes disruptive – urge is the now rather frequent use of instrumental recitative and arioso, such as the memorable 'beklemmt' ('constricted') passage in the Cavatina of the Quartet in B♭ op.130. Here instrumental music seems painfully to strive for articulate communication.

10. Ludwig van Beethoven: stipple engraving (1814) by Blasius Höfel after Louis Letronne

Several of the late works contain variation movements of a new kind. Earlier Beethoven had written
many brilliant piano variations, from the precocious
'Venni amore' set of 1790–91 to the C minor Variations
of 1806 – a series now to be capped by the encyclopedic
Thirty-three Variations on a Waltz by Diabelli. In his
first Vienna period, however, important variation
movements within larger works are not frequent. More
of these occur in the middle period. Generally the
variations are of the progressively decorative variety
(opp.57, 61, 67, 74, 97), a type that also continues into
the late years (opp.111, 125). But in the Sonata in E
op.109 and the late quartets, as well as in the Diabelli
set, Beethoven evolved a new type of variation in which
the members take a much more individual and profoundly reinterpreted view of the original theme. The
theme seems transformed or probed to its fundamentals, rather than merely varied. All this suggests a
changing concept of musical unity, now seen as an
evolution from within rather than as a conciliation of
contrasting forces: a Darwinian concept, perhaps,
rather than a Hegelian one.

In the most general sense, variation may also be said
to inspire the transcendent fugal finales of the Hammerklavier Sonata and the Quartet in Bb (the 'Grosse
Fuge'). The fugatos that occur in not a few of
Beethoven's earlier pieces hardly prepare one for his
preoccupation with contrapuntal forms in the late
years; scarcely a significant work now lacks an impressive fugal section or even a full-scale fugue bristling
with learned devices. Evidently he was looking for
some other means of musical movement than that

provided by the style he had inherited from Haydn and Mozart; fugue is a more dense, even style which places harmonic action in a very different light. In Beethoven's hands fugue became a means of flattening out the dramatic aspects of tonality. (It was not the only means that he devised, as witness the second and third movements of the Quartet in A minor.) Related to this general tendency is Beethoven's frequent avoidance in the late music of obvious dominant effects, his characteristic undercutting of tonic triads by 6-4 chords, and his somewhat wayward experiments with the church modes. As noted above, his early plans for a ninth symphony include a 'pious song ... in the ancient modes'.

There is in fact a persistent retrospective current in Beethoven's late period. He published or considered publishing several of his old songs (*Bundeslied, Der Kuss, Mit Mädeln sich vertragen*), reworked the *Opferlied* of 1794, resuscitated some old piano bagatelles for op.119 and reworked another in the second movement of the A minor Quartet. He finally set Schiller's *Ode to Joy* – a project first considered about 1790 – to a tune adumbrated in works of 1795 and 1808 (WoO 118, op.80). An archaizing urge is manifest in his interest in strict counterpoint and modality, even if the resulting pieces hardly sound archaic; over and above this, some of them look back to certain specific academic exercises recommended by Beethoven's old teacher Albrechtsberger. It was only in his late years that Beethoven developed his well-known penchant for writing canons *d'occasion*. Whereas in the 1800s he had spoken well of Cherubini, now his interest settled on Palestrina, Bach

– he sketched an overture on the notes B–A–C–H – and especially Handel. Handel's influence on the overture *Die Weihe des Hauses* (1822) is startling.

Yet ultimately Beethoven's real concern with fugue, as with variation and lyricism, was to mould these elements so that they could be embedded integrally into the matrix of the sonata style. The presentation, development and return of musical material within a finely controlled tonal field remained central to his artistic endeavour. Fugues perform the function of development sections in opp.101, 106, 111 and less directly in op.110 and the Ninth Symphony finale. Then the fugue at the beginning of the C♯ minor Quartet acts as an exposition, presenting the basic tonal and thematic material that is worked out in the rest of the piece. The variation movements of opp.127 and 135 have a powerful tonal dynamic built in. So does the Diabelli Variations – thanks to another fugue, which precedes the final, recapitulatory variation. Even *An die ferne Geliebte* arranges its cycle of six artless melodies in a purposeful order of keys and features a recapitulation followed by a miniature 'symphonic' coda.

VII   Late-period works
In some ways the few compositions finished between 1814 and 1816 – the song cycle and the sonatas op.90, op.102 nos.1 and 2, and op.101 – stand closer to Romantic music of the 1830s than any other Beethoven pieces. The opening movement of op.101, a genuine miniature sonata form in an unbroken lyrical sweep, begins quietly on the dominant as though the music

was already in progress: an almost Schumannesque effect. The returns of the first-movement themes (marked 'mit der innigsten Empfindung' and 'teneramente') later in the course of this sonata and in op.102 no.1 do not sound like characteristic Beethovenian recapitulations. They are nostalgic recollections which again suggest Schumann and his generation. All four sonatas carry on much further than before Beethoven's search for more fluid solutions to the problem of the form of the total sonata, in terms of the weight, balance and mood of the various movements.

The Sonata in B♭ op.106, arbitrarily, but not inappropriately called the 'Hammerklavier' (both op.101 and op.109 are also sub-titled 'für das Hammerklavier'), occupied Beethoven from late 1817 to late 1818; it was his first really large project in five years. Like the Eroica Symphony, it occupies a pivotal position in his output, though the differences between the two works are striking. The Eroica is one of the most popular and 'available' of his compositions, while the Hammerklavier is probably the most arcane. In different ways each represented a breakthrough for Beethoven, one like the crest of a great wave and the other like the breaking of a dam. And while both were works of revolutionary novelty, the Hammerklavier also paradoxically represents a reaction, in that Beethoven reverted to the traditional four-movement pattern in place of the fluid formal experiments of the sonatas of 1814–16, and turned away from their tone of lyrical intimacy.

One feature Beethoven did pick up from them was the idea of an abrasive fugal finale, present in the Cello

127

Sonata in D op.102 no.2. The Hammerklavier fugue with its famous *cancrizans* section is integrated into the total conception with astonishing care and rigour. An improvisatory introduction to the finale seems to grope for the fugue or, perhaps, to will it into existence (something of the kind happens in other late finales: opp.110, 125, 133, 135). Then the shape of the subject and the modulation plan both follow a pattern that has been established firmly (not to say exhaustively) in each of the previous movements. This is construction by means of descending 3rds, acting to a large extent as a substitute for the traditional dominant relation, and creating a superbly fruitful large-scale conflict between the tonic B♭ and B♮.

Beethoven had never written a work that depended so thoroughly, in all its aspects, on a single musical idea. The extremity of its conception, and of its demands on the performer, are as much a part of the character of this piece as are ideas of heroism in the Eroica.

In the three sonatas of 1820–22 Beethoven returned to the proportions and preoccupations of the sonatas of 1814–16. Of all his works, the Sonata in E op.109 is perhaps the most original in form, in respect both to its first movement and to the total aggregate. The first movement is another sonata form in an unbroken lyrical sweep, like the first movement of op.101, but much more complex and shadowy in quality, thanks first of all to the change from Vivace, ma non troppo to Adagio espressivo at the second group – after a mere eight bars. The next movement, an explosive Prestissimo, combines the functions of a more lucid sonata-form statement and a scherzo. A slow theme and variations follows, concluding with an extraordinarily serene unaltered da capo of the original hymn-like theme.

Under the lyrical spell of the Sonata in A♭ op.110, even the fugue in the finale is tuneful and positively smooth in counterpoint (it became so, at least, thanks to a terrific bout of sketching). And in the Sonata in C minor op.111, after the first movement has recalled in a spirtualized way all the 'C minor' gestures of the early Vienna years, the variations of the second (and last) movement create a visionary aura that had never been known in music before. This mood is recaptured at the very end of the Diabelli Variations.

Between October 1822 and February 1824 Beethoven completed three works which are in one way or another as gigantic as the Hammerklavier Sonata: the Diabelli Variations, the Mass in D (*Missa solemnis*) and the Ninth Symphony. Work on the variations and the mass had been in progress since early 1819. Beethoven's slowness in composing the mass can be explained in part by his inevitable resolve to approach the text in the highest seriousness and treat the setting as a personal testament. Indeed, the religious impetus spilled over into his next composition, the Ninth Symphony with its setting of stanzas from Schiller's half-bacchanalian, half-religious *Ode to Joy*. Mass and symphony stand together as the crowning statement about non-musical ideas in Beethoven's later life – a 'religious' statement to match or, rather, to supplant the 'heroic' statement made in the Eroica Symphony and *Leonore* nearly 20 years earlier. Between the two

129

late works there are many striking parallels of musical gesture and language.

But whereas the Ninth Symphony, despite grumblings that are heard from time to time about the finale, has always been and remains one of Beethoven's most successful and influential compositions, the same cannot be said of the Mass. It is perhaps unfortunate for the dissemination and appreciation of this magnificent work that the relaxed concert conventions of Beethoven's day – at the première only three separate movements were ventured – no longer obtain. If they did, the musical public might well come to appreciate and love the simpler movements, at least: the restrained and lyrical Kyrie, one of the composer's loveliest inventions; the Sanctus, with its organ-like interlude and ethereal violin solo in the Benedictus; and the Agnus, whose touching plea for what Beethoven described as 'inner and outer peace' is twice interrupted by exciting military fanfares and melodramatic recitatives, not to speak of one giddy modulating fugue.

Even the few statements made above are enough to suggest how much of this mass is unorthodox, both musically and liturgically. Unorthodoxies are multiplied in the Gloria and Credo (always the problematic movements for composers of masses). It is particularly in these two central movements that the traditions of the Viennese mass are made to accommodate older traditions deliberately resuscitated; Beethoven rubs shoulders with Haydn (the Haydn of the masses), Palestrina, Handel and Bach. Sublimity, awe and pathos are evoked unforgettably, but they are perhaps evoked too frequently and in too rapid a succession to

130

leave a satisfactory total impression. One can feel this even while acknowledging Beethoven's strenuous efforts at organization: the use of recurring themes for 'Gloria in excelsis Deo' and 'Credo, credo', the powerful tonal dynamic, and the weighting effect of the tremendous fugues 'In gloria Dei Patris' and 'Et vitam venturi'.

In the Mass Beethoven was obviously constrained by the pre-set text; in the symphony he chose his own text. He also chose the context for it: not within an intellectual liturgical structure, but in the real world of experience – for paradoxically or not, that is what the three opening instrumental movements meant to Beethoven. Ultimately, in the introduction to the finale, this world is rejected in favour of Schiller's transcendent vision of the joys of brotherhood, set 'without the adornments of Art' as an unaccompanied melody of universal folklike simplicity. From his experience of oneness with his fellows and with nature, says Schiller, man receives his intimation of a loving Father dwelling above the stars. This passage of the poem Beethoven set in a solemn religious style recalling that used in parts of the Mass in D.

As the one late-period Beethoven symphony, the Ninth is in a sense retrospective in resuming the 'symphonic' ideal which for a decade had inspired little music. Retrospective, too, is the frank echo of revolutionary French cantatas in the choral finale. Yet as a gesture, this finale shows once again Beethoven's uncanny grasp of essences below 'the adornments of Art'. As Wagner always insisted, words and a choir with soloists to sing them seem to force their way into

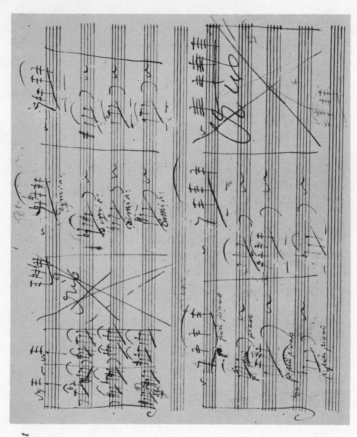

*11. Autograph MS of
part of the 5th movement
of String Quartet in A
minor, composed 1825*

the symphony in order to make instrumental music
fully articulate, to resolve the conflict of the earlier
movements with a consummation of unexampled
ecstasy.

In the late period Beethoven's treatment of sonata
form grows more and more subtle and even equivocal.
For example, he now tended to minimize the formal
development section and place a major climax after,
not at, the point of recapitulation (see opp.106, 130,
132). In the face of this, the first movement of the Ninth
provides a magnificent reassertion of the traditional
dynamic – though with a difference. During the famous
and much imitated introduction, the main theme
(another 'evolving' theme, one which seems to evolve
out of timeless infinity) grows up over a hollow
dominant 5th, A–E; then at the recapitulation this
returns *fortissimo* as a tonic D–A with F♯ in the bass:
an enhanced recapitulation from which all sense of
bluster has been filtered away and replaced by what one
can only call, with Tovey, catastrophe. The subsidiary
tonal areas of this movement, B♭ and a momentary B♮,
are 'composed out' in memorable fashion throughout
the rest of the symphony, as is the basic D minor/
major tropism of the first-movement recapitulation.

After completing the Ninth Symphony in early 1824,
Beethoven spent the two and a half years that remained
to him writing with increasing ease, it seems, and
exclusively in the medium of the string quartet. The five
late string quartets contain Beethoven's greatest music,
or so at least many listeners in the 20th century have
come to feel. The first of the five, op.127 in E♭ of
1823–4, shows all the important characteristics of this

unique body of music. It opens with another lyrical sonata form containing themes in two different tempos (as in op.109); the Maestoso theme melts into a faster one, wonderfully intimate and tender – even though it is constructed in three-part species counterpoint over a cantus firmus. The slow variation movement is of the new, more integral kind and the scherzo takes its impetus from a fugato. The finale burgeons with country-dance tunes, of a kind associated in the other late quartets with the interior dance movements (which one can scarcely call scherzos; certainly Beethoven no longer did so). In a brilliant coda, this finale submits to a sort of spirtualized dissolution, an effect prefigured in the Quartet in F minor op.95 and repeated in the next quartet, the A minor op.132.

The composition of op.132 was interrupted by a serious illness in April 1825, and an extraordinary 'Hymn of thanksgiving to the divinity, from a convalescent, in the Lydian mode' forms the central movement (of five). Beethoven's intimations of mortality take the form of modal cantus firmus variations dimly recalled from Albrechtsberger; they alternate movingly with a purely tonal section entitled 'Feeling new strength'. Cantus firmus writing is also in evidence in the first movement, as the themes in different tempos are now closely woven together. Extreme rhythmic fluidity combines with extreme concentration of detail. Beethoven had never before written such a deeply anguished composition.

In the Quartet in B♭ op.130, the confrontation of themes in different tempos gives the opening movement an elusive, even whimsical feeling. A deliberate sense of dissociation is intensified by the succession of five more movements, often in remote keys, with something of the effect of 'character pieces' in a Baroque suite. The feverish little Presto is followed by movements labelled by Beethoven Poco scherzando, Alla danza tedesca and Cavatina – and then by the 'Grosse Fuge', which seems to bear on its convulsive shoulders the responsibility for asserting order after so much disruption earlier in the piece. This it does by building its sections on various transformations of a cantus firmus subject – transformations closer in spirit to the Romantic symphonic poem than to any earlier fugal practice. The sections almost have the weight of separate movements, as in the Ninth Symphony finale. The lyric beauty of the slow G♭ section and the *Gemüthlichkeit* of the recurring section in 6/8 metre sometimes go unappreciated, it seems, by listeners awed by the determined dissonant fury of the others. A closed book to the 19th century, to Stravinsky the 'Grosse Fuge' was 'this absolutely contemporary piece of music that will be contemporary for ever'.

Years before, Beethoven had begun to extend the underlying principles of the sonata style to embrace the entire aggregate of movements in a piece. Now he found his largest movements breaking down into 'sub-movements' with a subsidiary integrity of their own. In the event, the Quarter in B♭ proved to be quite literally disruptive. In an unprecedented action, Beethoven sanctioned the removal of the fugal finale after the first performance, had it issued separately as op.133 and provided the quartet with a new, less radical (and less splendid) finale.

As though in reaction to this study in musical dissociation, Beethoven next wrote the most closely integrated of all his large compositions. From this point of view, the Quartet of C♯ minor op.131 may be seen as the culmination of his significant effort as a composer ever since going to Vienna. The seven movements run continuously into one another, and for the first time in Beethoven's music there is an emphatic and unmistakable thematic connection between the first movement and the last – not a reminiscence, but a functional parallel which helps bind the whole work together. A work of the deepest subtlety and beauty, at the end this quartet still seems to hinge on a stroke of the most elemental nature, as rushing D major scales in the finale recall the Neapolitan relationship set up between the opening fugue in C♯ minor and the following Allegro in D. Charles Rosen has remarked on Beethoven's continual

attempt to strip away, at some point in each large work, all decorative and even expressive elements from the musical material so that part of the structure of tonality is made to appear for a moment naked and immediate, and its presence in the rest of the work as a dynamic and temporal force suddenly becomes radiant.

A comparison with the analogous Neapolitan articulation at the end of the Quartet op.59 no.2 of 1806 shows how Beethoven could make such effects tell at the end of his lifetime.

The last quartet, op.135 in F, is a brilliant study in Classical nostalgia, though it does not lack a vision of the abyss in the second movement and a characteristic response through hymnody in the third. In the finale, when the main theme (marked with the words 'Es muss sein!') appears as a simple-minded inversion of the motif of the slow introduction (marked 'Muss es sein?'), a strong suspicion arises of parody – a self-parody of the familiar evolutionary slow introductions of these late years (cf op.111). The thematic tag itself was taken over from a contemporary humorous canon (WoO 196).

Like the Eighth Symphony, op.135 seems to mark the composer's farewell to a fully realized episode in his artistic journey. The writing of the late quartets was stimulated by external factors – Prince Golitsïn's commission and the return of the Schuppanzigh Quartet to Vienna – but it continued under its own impetus after the commission was fulfilled. The cohesiveness of this crowning episode of Beethoven's compositional activity is underlined by the observation made by various critics that three (or more) of the late quartets share melodic material. Even without this, they share some special stylistic characteristics; but even with all that, it is hard to accept the further implication that the individual works are aesthetically incomplete unless viewed as some sort of 'triptych' or 'cycle'. This may be true of the poems by T. S. Eliot which they inspired but not of the original quartets, any more than it is of other clearly associated works in Beethoven's output, such as the Mass in D and the Ninth Symphony.

CHAPTER THREE

Personal characteristics

Beethoven left an indelible impression on all those who encountered him in the years of his maturity, and even for his contemporaries there were certain features of his life – his idiosyncratic working methods, for example, his mournful isolation through deafness, and the nobility of his total dedication to his art – that endowed him as an almost mythical figure. The course subsequently taken by a romantic image of the composer in the years after his death is discussed in Chapter 4. Here something must be said of the realities from which the myths drew their strength.

He was neither good-looking nor equipped with more than a very rudimentary education; it was by the force of his character that he produced such a powerful effect on those around him. This, notoriously, had its thorny side. As a young man he was already known to be difficult, impatient and mistrustful, an 'unlicked bear'. A basic problem, it seems, was his ineptness at reading his own motives and interpreting those of others; thus misunderstandings were frequent, which his hot temper magnified into quarrels, even fisticuffs. But typically these were followed by reconciliations and scenes of penitence or remorse. What his capricious and at times outrageous behaviour could not dim was the enormous appeal of his personality. He fascinated and endeared himself to men and women of

138

many sorts, who continued to value his friendship no matter how rough a ride he gave them. This magnetic quality was most in evidence in his earlier years, but even near the end of his life, when he was often wretchedly ill and his deafness was impenetrable, there was competition for the privilege of rendering him services, and devoted friends were never far off.

In his relationships certain recurrent patterns can be observed. His male friendships fell into two broad types. There were the warm and intimate ones with companions such as Wegeler, Amenda and Stephan von Breuning, and perhaps also Franz von Brunsvik and Ignaz von Gleichenstein, men with whom he felt he could share his most private feelings and aspirations. Some considerable way behind came his relations with many others who were valued more for their disinterested usefulness to Beethoven than for any depth of shared emotion. Chief of these was the amiable bachelor Zmeskall; the two Lichnowksy brothers can also be counted among them, and in later years his factotum Schindler and perhaps the young Karl Holz, for whom however Beethoven also entertained some genuinely warm feelings. Most of the first select class of true friends were unmarried at the time of Beethoven's greatest intimacy with them. It is noteworthy, too, that both Wegeler and Amenda, the two with whom he maintained a serene relationship for the longest time, were in distant countries for most of his life; the friendship with Breuning, who remained in Vienna, was interrupted by a breach that lasted many years.

Beethoven's relations with women have been discussed much more fully than his friendships with men; they form the subject of a large but mainly speculative and sometimes very silly literature. He was certainly highly susceptible to feminine beauty and charm. The reliable Wegeler reported that 'he was never without a love, and most of them were from the upper ranks'. Of his attachments in Bonn little is known beyond a name or two, but in his early years in Vienna – again according to Wegeler – he was always involved in love affairs and 'made some conquests that many an Adonis would have found difficult if not impossible'. What these affairs amounted to is another matter. No doubt there were some trivial sexual adventures, but it is hard to avoid the impression that he also spent much time in a showy pursuit of women who could not, or would not, return his affection, and the very fact that most of them were 'from the upper ranks' meant that there was usually an insuperable barrier of social class to prevent the relationship from going too far. Though Beethoven always professed his desire for a true union of hearts, many of the women that he admired were contentedly married or were already committed to another man. Thus he was usually doomed to get nowhere – as perhaps, unconsciously, he intended. Something of the same pattern can be seen in the two or three relationships with women, described earlier, that involved Beethoven most deeply. To judge from the course that they took it seems plain that he shrank from a total involvement with a woman, and that he came to regard the household that he established with his nephew as in some ways a substitute for marriage.

That his life was in many respects lonely, therefore,

comes as no surprise. It is of course the overwhelming fact of his deafness that makes his personal history so poignantly different from that of other musicians. Its effect on his career was the long-term one of confirming the direction in which his interests were probably already turning; it obliged him at all events to commit himself almost entirely to composing, and to renounce any thoughts he may have had of pursuing fortune as a travelling virtuoso. But the impact of deafness on his social life was sharper and more immediate. It sank him in deep depression and led him to shun company for a time. In fact the years from 1800 to 1802, in which he brought himself to face the likelihood that his handicap would be permanent, were marked by a profound personal crisis, the resolution of which set the pattern for much of the rest of his life. Forced to recognize more and more that he was to be cut off from a part of human experience, he succeeded in coming to terms with an unusual and essentially solitary style of life. No doubt this reinforced his conviction, manifest even before the onset of deafness, that some of the rules of normal social behaviour did not apply to him.

There are many anecdotes of his peculiarities in this respect. Several concern his attitude to his superiors in rank, and to authority in general. Doubtless only too aware that he depended on aristocratic families for his financial support, he resolutely declined after his departure from Bonn to 'play the courtier' or to show the deference and obedience normally expected from musicians in circles of the nobility. He was often most unwilling, for instance, to perform on the piano if called on unexpectedly by his hosts to do so; sometimes he refused outright, and even left the soirée in a temper. He would also break off playing if people showed their inattention by chattering. The formal court etiquette that surrounded the Archduke Rudolph was especially irksome to him, and in the end it was Rudolph who surrendered by giving orders that the rules were not to be applied to Beethoven. Even in matters of dress Beethoven seems to have been unwilling to show the conformity expected of him, though in his earlier years in Vienna he was often smartly turned out.

This impatience with discipline and authority had more than one aspect. Temperamentally he was utterly unable to adopt a submissive attitude, and even in music he found it distasteful to accept the direction of living teachers (such as Haydn) or dead theoreticians. Moreover, as a child of his time, he was swayed by the ideals of the French Revolution; they must have dominated his student days, although a certain ambivalence can be detected in his attitude to them, as well as to the man who for a time embodied them, Napoleon. In his brusque dismissal of the conventions of an aristocratic society, in fact, Beethoven was less of the egalitarian than the élitist. He had little use for the common run of humanity, regarding himself as an artist – he was fond of the rather grand term 'Tondichter' ('poet in sound') – and, as such, at least the equal of anyone raised to eminence by birth or wealth. He accorded the greatest respect to other artists, particularly writers, and was puzzled and disappointed when he discovered that Goethe, whom he admired above all other poets, behaved over-deferentially to royal personages: was not Goethe as great as they were?

In matters of religion his views, as might be expected, were idiosyncratic and somewhat incoherent. It was not a subject that he discussed much with others. He was brought up in the tolerant Catholicism of the late 18th century, but the formal side of religion held little interest for him, though he went to some trouble while composing the *Missa solemnis* to ensure that he fully understood the words of the Mass. The deity of his faith was a personal God, a universal father to whom he constantly turned for consolation and forgiveness. That much is clear from the many private confessions and prayers scattered throughout his papers. Among philosophical books he was moved by the moral reflections of Kant. Perhaps more surprisingly, he found certain oriental writings on the immaterial nature of God sympathetic to him, and he copied out a number of their texts. He even framed some ancient Egyptian inscriptions on the nature of the deity and kept them on his writing-desk.

He also felt the presence of God in the beauty of nature, and sought to worship him in the countryside, having been greatly influenced by Christian Sturm's *Betrachtungen der Werke Gottes in der Natur*. Beethoven's love of the country was an enduring characteristic. He left Vienna for some months almost every summer and settled in one of the outlying villages such as Mödling or Heiligenstadt, or in the spa of Baden somewhat further afield. There he would take long, solitary walks in the woods and find refreshment of spirit. 'No-one', he wrote to Therese Malfatti in 1810, 'can love the country as much as I do. For surely woods, trees and rocks produce the echo which man desires to hear.'

When on his walks he would usually carry a bundle of folded music paper, and pause from time to time to make entries in it with a pencil. This activity was regarded by his contemporaries as a harmless eccentricity. They laughed too at his singular behaviour in restaurants, where he would sometimes sit for hours sunk in thought and then offer to pay for a meal that he had not eaten. Beethoven was certainly often strangely unaware of his physical surroundings and preoccupied with his own thoughts – even in Bonn the word 'raptus' had been jokingly applied to his fits of emotional inaccessibility – and the squalor of his rooms was such that only he could tolerate it.

Yet in relation to the thing about which he cared most – composing, and presenting his works to the public – Beethoven could hardly be said to be ill-organized. He had a regular domestic routine, rising early, making coffee by grinding a precise number of coffee-beans, and then working at his desk until two or three o'clock, when he had a meal. The morning's work was interrupted, though also in a sense maintained, by two or three short excursions out of doors, during which he continued to make sketches on music paper. Several of these 'pocket' sketchbooks have survived, together with a much larger number of 'desk' sketchbooks in which Beethoven worked when at home. The significance of these volumes, with page after page of seemingly illegible entries, was not understood by his contemporaries, who regarded his devotion to them as yet one more sign of his eccentricity. Only later did it come to be recognized that the sketchbooks provide a unique documentation, although a somewhat fragmentary and at times enigmatic one, of his creative processes.

When a work had been completed it was Beethoven's concern to find a publisher for it. The importance that he attached to publishers throughout his life is shown by the extent and range of his correspondence with them – for he contrived to persuade himself that his livelihood depended on selling his music to them, though he was in fact maintained largely by aristocratic subventions. At that time all a composer could expect was a lump sum for the sale of a work. Royalties were unknown. Nor was there any international copyright; within his own country a publisher usually enjoyed some protection for the works he had bought, but they could be freely copied (pirated) abroad. Thus it was a composer's concern to obtain the largest sum for each composition. In the case of Beethoven, whose later works involved many months or even years of labour, there was every inducement to compare the offers of various publishers and to play them off against each other – a form of behaviour that some modern critics, alerted no doubt by Beethoven's shrill protestations of commercial probity, have found unattractive.

One plan that interested him was that of publishing a work simultaneously in more than one country – something that Haydn had done with success. The advantage was that a composer could count on receiving two or more fees, and was thus able to settle for a lower sum from each publisher. From the publisher's point of view little was lost by sharing a work with a foreign publisher, since in practice the market of each country was more or less independent. In spite of the many practical difficulties of delivering manuscripts and synchronizing publication, Beethoven succeeded in getting a fair number of his compositions published by two or more firms in different countries at about the same time.

Beethoven's chief publishers may be briefly listed, together with the dates at which they were most active in publishing him: Artaria & Co., Vienna (1795–8), and two of Artaria's former partners, Tranquillo Mollo (1798–1801) and Giovanni Cappi (1802); Hoffmeister & Kühnel, Leipzig (1801–4); Bureau des Arts et d'Industrie, Vienna (1802–8); Breitkopf & Härtel, Leipzig (1802–3, 1809–12); Steiner & Co., Vienna (1815–17); A. M. Schlesinger, Berlin, and M. Schlesinger, Paris (1821–3 and 1827), Schott, Mainz (1825–7). Other publishers, such as Simrock of Bonn, occasionally issued important works. In the English market one firm predominated: that of Muzio Clementi (1810–23), who secured the English rights to a large number of works by direct dealings with Beethoven, and brought them out at the same time as the Viennese, Leipzig or Paris editions. Since these English editions were produced independently of the continental ones, each is potentially important for establishing an authentic text. George Thomson of Edinburgh also deserves a word or two. A civil servant and

12. Ludwig van Beethoven: drawing by Johann Peter Theodor Lyser (first published in 1833)

musical amateur who devoted much of his life to collecting national (and particularly Scottish) folk-songs, he had already published several volumes of melodies with accompaniments by Pleyel, Kozeluch and Haydn before he approached Beethoven in 1803 with the request that he should write six sonatas introducing Scottish melodies. Although nothing came of this suggestion or other similar ones, Beethoven did in the end undertake to write piano trio accompaniments to a great quantity of Scottish, Welsh and Irish melodies submitted by Thomson. The work was carried out between 1810 and 1818, a period that included some otherwise barren years. About 1818 Beethoven also wrote for Thomson some simple variations for flute and piano on national melodies (opp.105, 107).

Physically Beethoven was of no more than average height, but his stocky frame conveyed a sense of great muscular strength. He had broad shoulders and a short neck. His pockmarked face, with its wide nose and bushy eyebrows, was described by some as ugly and was certainly remote from the conventional good looks of the time, although it was recognized as having a quality of nobility about it. In youth his hair was coal-black and his complexion swarthy; in middle age, partly as a result of ill-health, his hair became grey and his face rather florid. What impressed those who met him was the intensity of the gaze from his deep-set eyes, and the enormous amimation of his melancholy features and indeed of the whole of his restless body. This vitality is not captured in most of the portraits and sketches made in his lifetime. The best representation is probably the 1814 engraving by Blasius Höfel (based on a pencil drawing by Louis Letronne, but touched up from the life; fig.10, p.123). The bust by Franz Klein (fig.4, p.59) is based on a life-mask of 1812, so the features have claims to authenticity; and the sketches by Lyser showing Beethoven walking in the street, though not authenticated, also carry conviction (fig.12, p.147). The idealized portraits and busts of more recent years must be regarded as part of the Beethoven cult; they owe nothing to literal or even to poetic truth.

148

CHAPTER FOUR

Posthumous influence and reputation

Whole books have been devoted to single aspects of Beethoven's posthumous reputation. The first and most striking element of this is the great and abiding popularity of his music. During the last years of his life and the period after his death the musical audience was changing, as a new bourgeois element replaced the typical 18th-century aristocratic circles for which he himself had written. To this new audience Beethoven's music appealed with particular (and almost uncanny) force. His symphonies, concertos, overtures and the more famous of his piano sonatas at once became central to the musical culture of the 19th century, and have remained so to the present, even as the character of the musical audience has changed again. Musicians, standing somewhat outside today's mass audience, now tend to reserve their greatest admiration for the music of Beethoven's third period. But for most of them this has not replaced the traditional central repertory; it has simply augmented it.

Beethoven's fame was and is, however, more than a purely musical phenomenon. The story of his life – outwardly so uneventful, yet so full of inner pathos – became inextricably blended with the particular qualities of his music to produce a composite image which fascinated the age of Romanticism and exerted a powerful, and sometimes baleful, effect on the careers of other musicians. More than any other composer, painter or author, Beethoven was felt to represent the very type of the artist – a figure that came to assume mythical proportions in the Romantic consciousness. Indeed, it is not too much to speak of a Beethoven myth, based on but rapidly outpacing biographical and musical realities.

This was generated principally by (and for) music lovers, amateurs, and popular and semi-popular writers on music. But professionals were also involved, and since so many 19th-century composers were also writers and propagandists, they cannot as a class be excluded from the ranks of the myth-makers.

They can be forgiven though, for few escaped his influence in technical respects. Of course this manifested itself in widely different ways. Schubert, only 26 years his junior, confessed himself puzzled and disturbed by some of Beethoven's compositions, but was in a position to learn from them more directly than any other of the great masters. Beethovenian models are apparent in each of the last three piano sonatas, for example, and even in some songs (*Mignon*, *Der Wachtelschlag*). The measured tread of the slow movement of the Seventh Symphony is heard echoing in Schubert's C major Symphony, Mendelssohn's 'Italian', and Berlioz's *Harold en Italie*, where in addition the rehearsal of themes at the beginning of the finale derives from Beethoven's Ninth. As for the very beginning of the Ninth, that resonates endlessly in a whole line of compositions by Bruckner, Mahler,

Richard Strauss and others: while Wagner never tired of pointing to the Choral Symphony as proof that music of the future must lie on a line of development from wordless symphony to music drama. The 'Moonlight' Sonata makes an appearance in a Druid chorus in Bellini's *Norma*. Schumann found in *An die ferne Geliebte* much to ponder – and something to quote. Brahms's Beethoven fixation is well known; Dvořák is inconceivable without the Pastoral Symphony. Schoenberg said that his First String Quartet was modelled on the Eroica, and more generally the late Beethoven quartets stand as an obvious inspiration for those of Bartók. It was perhaps only a generation ready to write 'Schoenberg is dead' on its banner that could finally free itself from Beethoven's influence.

The artist as hero had begun to haunt German Romantic literature before 1800 and was still doing so at the time of Hesse and Mann. E. T. A. Hoffmann, who popularized the idea of the 'daemonic' musician in his stories, also wrote adulatory (and remarkably penetrating) criticism of Beethoven compositions as early as 1810. Beethoven's heroism was seen in his force of character, his independence and libertarianism, his deafness and conspicuous suffering and failure to find a woman's love, his single-minded devotion to his art, and his manifest (as it seemed) ability to transform lonely adversity into a series of affirmative artistic visions. The Heiligenstadt Testament, the letter to the 'Immortal Beloved' and the imaginative memoirs of Bettina Brentano fed the myth; so did titles supplied by the composer himself (*Sinfonia eroica, Sonate pathétique*) or by others ('Appassionata', 'Emperor'); so did numerous anecdotes which were not always the less typical for being apocryphal. One told by Bettina Brentano, for example, was widely credited: when Beethoven was walking with Goethe in Teplitz in 1812 and royalty came passing by, Goethe doffed his hat and stood deferentially on one side while Beethoven strode on staring straight before him. 'My nobility is *here* and *here*', he is said to have exclaimed on another occasion, pointing to his heart and his head. To the 19th century, unaware of how highly Beethoven prized the 'van' in his name, this told of the inherent nobility of man's individuality and creativity over and above the accidents of privilege.

The Beethoven myth took history or historicism in its stride; from *Beethoven the Creator* it was only a short step to *Beethoven, Life of a Conqueror* and then to *Beethoven, the Man who Freed Music* (these are titles of books widely read in the early part of the 20th century). He was the revolutionary who shattered the mould of Classical music, and he was also ideally situated to support the ideological view of a Teutonic hegemony in great music, so dear to German musicians from the time of Schumann on. This theme was sounded most insistently by Wagner, whose important monograph *Beethoven* appeared in 1870, the centenary of the composer's birth and the year of the Franco-Prussian War.

In this year, also, the house where Beethoven was born was made into a national shrine beside the River Rhine (later an important manuscript archive and research institute were added to the Beethovenhaus

13. Ludwig van Beethoven: statue (1886–1902) by Max Klinger

organization). If the foundation of the Bonn Beethovenhaus can be viewed as pressing the claims of the country of Beethoven's birth over that of his adoption, Vienna responded in 1900 by unveiling the most grandiose and also the most distinguished of all the many monuments raised by the 19th century to its hero. This was an extraordinary nude statue by Klinger (fig.13) placed in a room decorated by Klimt with scenes (which have not survived) from Beethoven's compositions. At the dedication ceremonies Mahler conducted a brass band and a massed choir of Viennese working men in an outdoor performance of the *Ode to Joy*.

153

Each in its own way, the two greatest monuments of
Beethoven scholarship were also closely bound to the
Beethoven myth. The exemplary biography by Thayer
could scarcely have been written by a contemporary
German, or perhaps by any contemporary European.
Fired by a passion for strict objectivity matched only by
his admiration for Beethoven, Thayer devoted a lifetime to the most painstaking sifting of 'hard' evidence in order to correct the mass of misinformation
that had grown up around his subject and to debunk
romantic inventions. Yet, ironically enough, Thayer's
very American programme of objective inquiry un-
covered rather more than he had counted on. Growing
awareness of certain aspects of Beethoven's character
went so sharply against his own Victorian preconceptions, apparently, that he found it impossible to
proceed with the study, and he never completed it.

The other monument is the comprehensive survey of
and commentary on Beethoven's copious sketches by
Gustav Nottebohm, a Viennese scholar and ped-
agogue, pupil of Schumann and Mendelssohn and
close friend of Brahms. Nottebohm's discoveries in the
sketchbooks provided remarkable insight into the
inner workings of the music; few works of musicology
have been received so warmly by the musical world at
large. And they also enhanced the Beethoven myth
significantly, by making graphic the labour and devotion that he accorded to every act of creation. The
Victorian era was quick to admire the heroic regime of
self-improvement which he could be seen to have
followed, and the Darwinians pored over the evidence
of great masterpieces growing step by step from the

simplest musical seeds. Nottebohm's researches were resumed in a systematic way only in the 1960s, mainly by musicologists from the USA and Britain, and led somewhat belatedly to a wave of scholarly interest in the sketches of other composers.

In addition to the works of Thayer and Nottebohm, the 19th century produced some of its most impressive analytical and critical writing in response to music by Beethoven. As Dahlhaus has emphasized, those same famous Hoffmann reviews that project the idea of 'absolute music' – especially Beethoven's – as a direct route into the spirit realm also include lengthy technical analyses, through which Hoffmann meant to support his idea by revealing the music's autonomous organic perfection. Less metaphysically he was followed by Lenz (1852), Marx (1859, 1863), Berlioz (1862), Grove (1884) and Helm (1885). No other composer stimulated such an array of technical literature, with the result that Beethoven was elevated to a position alongside Bach as the paragon of academic music study. Then in the decades around 1900, when the great tradition of the 'Three B's' was seen to be in danger, his music became central to the evolving thought of theorists who are still of commanding importance in the 1970s. Schenker, Tovey and Réti were to devote their important studies to Beethoven: Schenker's monographs on the Third, Fifth and Ninth Symphonies and the late piano sonatas, Tovey's *Companion to Beethoven's Pianoforte Sonatas* and *Beethoven*, and Réti's *Thematic Patterns in the Sonatas of Beethoven*.

The early 20th century also produced a fitting climax to Beethoven hagiography in *Jean-Christophe* (1904–12), an exceedingly long and popular novel based on a romanticized view of the composer's life by the universal man of letters and musicologist Romain Rolland. In the reaction against Romanticism after World War I some voices (Stravinsky, Dent) were raised against Beethoven, particularly the Beethoven of the 'C minor mood' and the famous symphonies, and there were even those who posed sceptical questions about his personality (Newman). But perhaps exactly because Beethoven was not a Romantic artist, or not purely a Romantic artist – commentators differ sharply on just how this should be formulated – such antipathy did not spread far. Leading performing musicians like Toscanini, Schnabel and the Budapest Quartet built careers round their Beethoven performances, and the later 20th century was able to accept such revisionist studies as *Beethoven and his Nephew* by the psychoanalysts E. and R. Sterba – over-hostile, but sobering – and the more sympathetic, balanced, and comprehensive psychobiography by Maynard Solomon.

Beethoven has survived demythification. At the second centenary of his birth his popularity and reputation seemed undiminished. The anniversary years 1970 and 1977 were celebrated by, among other things, scholarly conferences in Bonn, East Berlin, Vienna and Chapel Hill, and by the issue of two 'complete' recordings of his music, on 77 and 111 long-playing records. More significantly, in the 1970s and 1980s performances and recordings of his music on original instruments, notably the fortepiano, have provided a new dimension of insight into the seemingly inexhaustible fund of Beethoven's art. It is likely that future generations will find still more.

WORKS

Editions: L. van Beethoven: Werke: Vollständige kritisch durchgesehene überall berechtigte Ausgabe, i–xxiv (Leipzig, 1862–5), xxv [suppl.] (Leipzig, 1888) [GA]

L. van Beethoven: Sämtliche Werke: Supplemente zur Gesamtausgabe, ed. W. Hess (Wiesbaden, 1959–71) [HS]

L. van Beethoven: Werke: neue Ausgabe sämtlicher Werke, ed. J. Schmidt-Görg and others (Munich and Duisburg, 1961–) [NA]

Works are identified in the left-hand column by opus and WoO (Werk ohne Opuszahl, 'work without opus number') numbers as listed in G. Kinsky and H. Halm: Das Werk Beethovens (Munich and Duisburg, 1955) and by Hess numbers as listed in W. Hess: Verzeichnis der nicht in der Gesamtausgabe veröffentlichten Werke Ludwig van Beethovens (Wiesbaden, 1957). Works published in GA are identified by the volume in which they appear (roman numeral) and the position in the publisher's continuous numeration (arabic number); works published in HS are listed in the GA column and identified by volume number. Works published in NA are identified by category (roman numeral) and volume within each category (arabic number).

p – parts   s – full score   vs – vocal score

Numbers in the right-hand column denote references in the text.

ORCHESTRAL

No.	Title, Key	Composition, First performance	Publication	Dedication, Remarks	GA	NA	
op.21	Symphony no.1, C	1799–1800; 2 April 1800	p: Leipzig, 1801	Baron Gottfried van Swieten	i/1		27, 29, 36, 99, 104, 107
op.36	Symphony no.2, D	1801–2; 5 April 1803	p: Vienna, 1804	Prince Karl von Lichnowsky	i/2		34, 35, 104, 105, 107
op.55	Symphony no.3 'Eroica', Eb	1803; 7 April 1805	p: Vienna, 1806	Prince Franz Joseph von Lobkowitz	i/3		37, 39, 91, 102, 104, 105ff, 127, 151
op.60	Symphony no.4, Bb	1806; March 1807	p: Vienna, 1808	Count Franz von Oppersdorff	i/4		43, 111
op.138	Overture 'Leonore no.1', C	1806–7; 7 Feb 1828	s, p: Vienna, 1838	for Leonore ovs. nos.2–3, see 'Operas'	iii/19		44
op.62	Overture to Collin's Coriolan, c	1807; March 1807	p: Vienna, 1808	Heinrich Joseph von Collin	iii/18	ii/1	44, 98
op.67	Symphony no.5, c	1807–8; 22 Dec 1808	p: Leipzig, 1809	Prince Lobkowitz and Count Andreas Razumovsky; preliminary sketches, 1804	i/5		45, 46, 98, 107, 108, 110, 124
op.68	Symphony no.6 'Pastoral', F	1808; 22 Dec 1808	p: Leipzig, 1809	Prince Lobkowitz and Count Razumovsky	i/6		45, 46, 107f, 111f, 112, 151

158

No.	Title, Key	Composition, First performance	Publication	Dedication, Remarks	GA	NA	
op.92	Symphony no.7, A	1811–12; 8 Dec 1813	s, p: Vienna, 1816	Count Moritz von Fries; arrs. for pf, pf 4 hands and 2 pf; ded. Elisabeth Alexiewna, Empress of Russia	i/7		52, 58, 62, 107, 108, 118f, 150
op.93	Symphony no.8, F	1812; 27 Feb 1814	s, p: Vienna, 1817	shortened version of end of 1st movt, HS iv	i/8		52, 53, 58, 119
op.91	Wellingtons Sieg oder Die Schlacht bei Vittoria ('Battle Symphony')	1813; 8 Dec 1813	s, p: Vienna, 1816; for pf: London and Vienna, 1816	Prince Regent of England (later King George IV); orig. version of pt.2, for Maelzel's panharmonicon, HS iv	ii/10; HS viii [for pf]	ii/1	58, 62, 68, 91
op.115	Overture 'Namensfeier', C	1814–15; 25 Dec 1815	s, p: Vienna, 1825	Prince Anton Heinrich Radziwill; incorporates ideas sketched several years earlier	iii/22	ii/1	61, 75
WoO 3	Gratulations-Menuet, Eb	1822; 3 Nov 1822	p: Vienna, 1832	written for Carl Friedrich Hensler, ded. (by publisher) Karl Holz	ii/13	ii/3	
op.125	Symphony no.9, d	1822–4; 7 May 1824	s, p: Mainz, 1826	Friedrich Wilhelm III of Prussia	i/9		74, 76, 77, 78, 79, 107, 122, 124, 126, 128, 129, 130f, 135, 137, 150

SOLO INSTRUMENTS AND ORCHESTRA

No.	Title, Key	Composition, First performance	Publication	Dedication, Remarks	GA	NA
WoO 4	Piano Concerto, Eb	1784	s: GA	survives only in pf score (with orch cues in solo part)	xxv/310	
Hess 13	Romance, e, pf, fl, bn, orch, frag.	1786	Wiesbaden, 1952	intended as slow movt of larger work	HS iii	
WoO 5	Violin Concerto, C, frag.	c1790–92	Vienna, 1879	part of 1st movt only; 1st edn. ded. Gerhard von Breuning	HS iii	
Hess 12	Oboe Concerto, F, lost	?1792–3	—	sent to Bonn from Vienna in late 1793; a few sketches survive	—	—

No	Title, Key	Composition, First performance	Publication	Dedication, Remarks	GA	NA	
WoO 6	Rondo, Bb, pf, orch	before 1794	p: Vienna, 1829	orig. finale of op.19; solo part completed by Czerny for 1st edn.	ix/72; HS iii		
op.19	Piano Concerto no.2, Bb	begun before 1793, rev. 1794–5, 1798; 29 March 1795	p: Leipzig, 1801	Carl Nicklas von Nickelsberg; score frag. rejected from early version, HS iii	ix/66		19, 24, 27, 29, 99
	cadenza for 1st movt	?1809	GA		ix/70a		
op.15	Piano Concerto no.1, C	1795, ? rev. 1800; 18 Dec 1795	p: Vienna, 1801	Princess Barbara Odescalchi (née Countess von Keglevics)	ix/65		19, 20, 24, 27
op.50	3 cadenzas for 1st movt Romance, F, vn, orch	?1809 ?1798; ? Nov 1798	GA p: Vienna, 1805		ix/70a iv/31	iii/4	
op.37	Piano Concerto no.3, c	?1800; 5 April 1803	p: Vienna, 1804	Prince Louis Ferdinand of Prussia; ? rev. 1803	ix/67		27, 36, 98
	cadenza for 1st movt	?1809	GA		ix/70a		
op.40	Romance, G, vn, orch	?1801–2	p: Leipzig, 1803		iv/30	iii/4	
op.56	Concerto ('Triple Concerto'), C, pf, vn, vc, orch	1803–4; May 1808	p: Vienna, 1807	Prince Lobkowitz	ix/70	iii/1	
op.58	Piano Concerto no.4, G	1805–6; March 1807	p: Vienna, 1808	Archduke Rudolph of Austria	ix/68		43, 46, 118
	2 cadenzas for 1st movt, cadenza for finale	?1809	GA		ix/70a		
	cadenza for 1st movt, 2 cadenzas for finale (Hess 81, 82, 83)	?1809	NA		HS x	vii/7	
op.61	Violin Concerto, D	1806; 23 Dec 1806	p: Vienna, 1808; London, 1810	Stephan von Breuning	iv/29; HS x	iii/4	43, 124
—	arr. of Vn Conc. op.61 as a pf conc.	1807	p: Vienna, 1808; London, 1810	Julie von Breuning	ix/73 (solo pt)		
	cadenza for 1st movt, cadenza for finale	?1809	GA		ix/70a		
	2 cadenzas for finale (Hess 84–5)	?1809	NA		HS x	vii/7	

No.	Title, Key	Composition	Publication	Dedication. Remarks	GA	NA	
op.80	Fantasia, c, pf, chorus, orch ('Choral Fantasy')	1808; 22 Dec 1808	p: London, 1811; Leipzig, 1811	Maximilian Joseph, King of Bavaria; notation of solo part completed 1809; string parts for rejected orch introduction, HS x	ix/71		
op.73	Piano Concerto no.5 'Emperor', Eb	1809; ?28 Nov 1811	p: London, 1810; Leipzig, 1811	Archduke Rudolph	ix/69		

WIND BAND

No.	Title, Key	Composition	Publication	Dedication. Remarks	GA	NA	
WoO 29	March, Bb, 2 cl, 2 hn, 2 bn: see 'Chamber Music for Wind alone and with Strings'						
WoO 18	March 'für die böhmische Landwehr', F	1809	pf red.: Prague, ?1809; s: Berlin, ?1818–19	Archduke Anton of Austria (on autograph)	xxv/287/1; HS iv; HS viii [pf]		
	trio to WoO 18, Bb	?1822–3	HS		HS iv		
WoO 19	March, F	1810	pf red.: Vienna, 1810	Archduke Anton (on autograph)	xxv/287/2; HS iv		
	trio to WoO 19, f	?1822–3	HS				
WoO 21	Polonaise, D	1810	GA		xxv/289		
WoO 22	Ecossaise, D	1810	GA		xxv/290		
WoO 23	Ecossaise, G	?1810	pf red. in Czerny's Musikalisches Pfennig-Magazin, i (Vienna, 1834)		xxv/306 [pf]		69
WoO 24	March, D	1816	pf red.: Vienna, 1827		ii/15		
WoO 20	March with Trio, C	before 1823	GA		xxv/288; HS iv		

161

CHAMBER MUSIC FOR STRINGS

No.	Title, Key	Composition	Publication	Dedication, Remarks	GA	NA	
Hess 33	Minuet, A♭, str qt	c1790	HS	exists also in pf version, HS viii	HS vi		
op.3	String Trio, E♭	before 1794	Vienna, 1796	frag. arr. Beethoven for pf trio, HS ix	vii/54; HS vi	vi/6	92, 95
op.87	transcr. of Trio for 2 ob and eng hn op.87 as str trio, C	1795	Vienna, 1806	transcr. probably approved by Beethoven	—	vi/6	
op.4	String Quintet, E♭	1795	Vienna, 1796	thoroughly recomposed version of Octet op.103 (see 'Chamber Music for Wind alone and with Strings')	v/36	vi/2	
WoO 32	Duet, E♭, va, vc 'mit zwei obligaten Augengläsern', 1st mcvt and minuet	1796-7	1st movt: Leipzig, 1912; minuet: Frankfurt am Main, London and New York, 1952	probably written for Nikolaus Zmeskall von Domanovecz	HS vi	vi/6	
op.8	Serenade, D, str, trio	1796-7	Vienna, 1797		vii/58	vi/6	
op.9	Three String Trios, G, D, c	1797-8	Vienna, 1798	Count Johann Georg von Browne	vii/55-7	vi/6	23, 98, 99
Hess 28	another trio for the minuet of op.9 no.1, G	1797-8	Bonn, 1924		HS vi	vi/6	
op.18	Six String Quartets, F, G, D, c, A, B♭	1798-1800	Vienna, 1801	Prince Lobkowitz; early version of no.1, HS vi, NA	vi/37-42	vi/3	26, 28, 92, 98, 99, 100, 102
op.29	String Quintet, C	1801	Leipzig, 1802	Count Fries	v/34	vi/2	
Hess 34	arr. of Pf Sonata op.14 no.1 for str qt, F	1801-2	Vienna, 1802	Baroness Josefine von Braun	HS vi	vi/3	104
op.59	Three String Quartets 'Razumovsky', F, e, C	1805-6	Vienna, 1808	Count Razumovsky	vi/43-5	vi/4	37, 114, 118, 119, 136
op.74	String Quartet 'Harp', E♭	1809	Leipzig and London, 1810	Prince Lobkowitz	vi/46	vi/4	50, 118, 124
op.95	String Quartet 'Serioso', f	1810	Vienna, 1816	Zmeskall von Domanovecz	vi/47	vi/4	52, 119, 134
op.104	arr. of Pf Trio op.1 no.3 for str qnt	1817	Vienna and London, 1819	arr. corrected by Beethoven, but largely the work of someone else	v/36a	vi/2	69

					HS vi	vi/2
Hess 40	Prelude, d, str qnt	?1817	*SMz*, xcv (1959)		HS vi	vi/2
op.137	Fugue, D, str qnt	1817	Vienna, 1827		v/35	
WoO 34	Duet, A, 2 vn	1822	T. von Frimmel: *Ludwig van Beethoven* (Berlin, 1901)	Alexandre Boucher	HS vi	
op.127	String Quartet, E♭	1823–5	Mainz, 1826	Prince Nikolai Golitsïn	vi/48	80, 121, 126, 133
op.132	String Quartet, a	1825	Paris and Berlin, 1827	Prince Golitsïn	vi/51	81, 82, 125, 132, 133, 134
op.130	String Quartet, B♭	1825–6	Vienna, 1827	Prince Golitsïn; orig. with op.133 as finale; new finale composed 1826	vi/49	82, 86, 111, 122, 133, 134
op.133	Grosse Fuge, B♭, str qt	1825–6	Vienna, 1827	Archduke Rudolph; orig. finale of op.130	vi/53	86, 124, 128, 135
op.131	String Quartet, c♯	1826	Mainz, 1827	Baron Joseph von Stutterheim	vi/50	82, 85, 108, 126, 136
op.135	String Quartet, F	1826	Berlin and Paris, 1827	Johann Wolfmayer	vi/52	86, 126, 128, 136, 137
Hess 41	String Quintet, C, frag.	1826	Vienna, 1838	survives only in pf transcr., WoO 62	HS viii	

CHAMBER MUSIC FOR WIND ALONE AND WITH STRINGS

WoO 26	Allegro and Minuet, G, 2 fl	1792	A. W. Thayer: *Ludwig van Beethovens Leben*, ed. H. Deiters, ii (Berlin, 1901)	J. M. Degenhart	HS vii	
op.103	Octet, E♭, 2 ob, 2 cl, 2 hn, 2 bn	?1792–3	Vienna, 1830	probably written before 1793, then rev. 1793	viii/59	94, 95
WoO 25	Rondino, E♭, 2 ob, 2 cl, 2 hn, 2 bn	1793	Vienna, 1830	at one time intended as finale to op.103	viii/60	
Hess 19	Quintet, E♭, ob, 3 hn, bn, inc.	?1793	Mainz, 1954	incl. 1st movt frag., slow movt, minuet frag.; probably begun before 1793, then rev. 1793	HS vii	
op.87	Trio, C, 2 ob, eng hn	1795	Vienna, 1806		viii/63	

No.	Title, Key	Composition	Publication	Dedication, Remarks	GA	NA	
WoO 28	Variations, C, on 'La ci darem la mano' from Don Giovanni, 2 ob, eng hn	?1795	Leipzig, 1914		HS vii		
op.81b	Sextet, E♭, 2 hn, 2 vn, va, vc	?1795	Bonn, 1810		v/33		
op.71	Sextet, E♭, 2 cl, 2 hn, 2 bn	1796	Leipzig, 1810	1st and 2nd movts probably written before 1796	viii/61		
WoO 29	March, B♭, 2 cl, 2 hn, 2 bn	1798	GA	pf version pubd in *Schweizerische musikpädagogische Blätter*, xx (1931), also HS viii; see also 'Miscellaneous', Hess 107	xxv/292		27, 29, 104
op.20	Septet, E♭, cl, hn, bn, vn, va, vc, db	1799–1800	Leipzig, 1802	Empress Maria Theresa	v/32		
op.25	Serenade, D, fl, vn, va	1801	Vienna, 1802		viii/62		
WoO 30	Three Equali, 4 trbn: d, D, B♭	1812	GA	transcr. for 4 male vv by I. von Seyfried perf. at Beethoven's funeral, pubd Vienna, 1827	xxv/293		
WoO 17	Eleven Dances ('Mödlinger Tänze'), wind, str: see 'Collections of Dances'						
		CHAMBER MUSIC WITH PIANO					
WoO 36	Three Quartets, pf, vn, va, b: E♭, D, C	1785	Vienna, 1828	autograph gives 'clavecin' instead of pf	x/75–7	iv/1	4
WoO 37	Trio, G, pf, fl, bn	1786	GA	autograph gives 'clavicembalo' instead of pf	xxv/294	iv/3	
WoO 38	Piano Trio, E♭	?1791	Frankfurt, 1830	date of composition taken from early catalogue of Beethoven's works	xi/86	iv/3	94
Hess 48	Allegretto, E♭, pf trio	c1790–92	London, 1955	authenticity no longer in doubt	HS ix	iv/3	
Hess 46	Violin Sonata, A, frag.	c1790–92	HS		HS ix		
WoO 40	Variations, F, on 'Se vuol ballare' from Le nozze di Figaro, pf, vn	1792–3	Vienna, 1793	Eleonore von Breuning	xii/103	v/2	18, 93

WoO 41	Rondo, G, pf, vn	1793–4	Bonn, 1808		xii/102		19, 69, 95, 96, 97
op.1	Three Piano Trios, Eb, G, c	1794–5	Vienna, 1795	Prince Lichnowsky; no.1 probably composed before 1794	xi/79–81	v/2	
WoO 43a	Sonatina, c, pf, mand	1796	*Grove 1* ('Mandoline')	probably written for Countess Josephine de Clary	xxv/295		
WoO 43b	Adagio, Eb, pf, mand	1796	GA	probably for Countess de Clary; a slightly variant version pubd in *Sudetendeutsches Musikarchiv* (1940), no.2	xxv/296; HS ix		
WoO 44a	Sonatina, C, pf, mand	1796	*Der Merker*, iii (1912)	probably for Countess de Clary	HS ix		
WoO 44b	Andante and Variations, D, pf, mand	1796	*Sudetendeutsches Musikarchiv* (940), no.1	probably for Countess de Clary	HS ix		
WoO 42	Six German Dances, pf, vn: see 'Collections of Dances'						
op.5	Two Cello Sonatas, F, g	1796	Vienna, 1797	Friedrich Wilhelm II of Prussia	xiii/105–6	v/3	21, 23, 99
WoO 45	Variations, G, on 'See the conqu'ring hero comes' from Judas Maccabaeus, pf, vc	1796	Vienna, 1797	Princess Christiane von Lichnowsky	xiii/110	v/3	21
op.66	Variations, F, on 'Ein Mädchen oder Weibchen' from Die Zauberflöte pf, vc	1796	Vienna, 1798		xiii/111	v/3	
op.16	Quintet, Eb, pf, ob, cl, hn, bn	1796	Vienna, 1801	Prince Joseph Johann zu Schwarzenberg	x/74	iv/1	92, 99
	arr. of op.16 for pf qt		Vienna 1801 (with pf and wind version)	Prince Schwarzenberg; authenticity affirmed in Wegeler and Ries (1838)	x/78	iv/1	
op.11	Trio, Bb, pf, cl/vn, vc	1797	Vienna, 1798	Countess Maria Wilhelmine von Thun	xi/89		
op.12	Three Violin Sonatas, D, A, Eb	1797–8	Vienna, 1799	Antonio Salieri	xii/92–4	v/1	99
op.17	Horn Sonata, F	1800	Vienna, 1801	Baroness Josefine von Braun	xiv/112		28

No.	Title, Key	Composition	Publication	Dedication, Remarks	GA	NA	
op.23	Violin Sonata, a	1800	Vienna, 1801	Count Fries	xii/95	v/1	28
op.24	Violin Sonata 'Spring', F	1800–01	Vienna, 1801	Count Fries	xii/96	v/1	28
WoO 46	Variations, E♭, on 'Bei Männern, welche Liebe fühlen' from Die Zauberflöte, pf, vc	1801	Vienna, 1802	Count von Browne	xiii/111a	v/3	
op.30	Three Violin Sonatas, A, c, G	1801–2	Vienna, 1803	Alexander I, Tsar of Russia	xii/97–9	v/2	34, 98, *101*
op.44	Variations, E♭, on an original theme, pf trio		Leipzig, 1804	sketched in 1792	xi/88	iv/3	
op.47	Violin Sonata 'Kreutzer', a	1802–3	Bonn and London, 1805	written for George P. Bridgetower, ded. Rodolphe Kreutzer; finale orig. intended for op.30/1	xii/100	v/2	37, 99, 113
op.38	Trio, E♭, pf, cl/vn, vc (arr. of Septet op.20: see 'Chamber Music for Wind alone and with Strings')	?1803	Vienna, 1805	Professor Johann Adam Schmidt	xi/91	iv/3	
op.121a	Variations, G, on Wenzel Müller's 'Ich bin der Schneider Kakadu', pf trio	?1803, rev. 1816	Vienna and London, 1824	probably offered for publication in 1803; surviving autograph dates from c1816–17	xi/87	iv/3	69
op.41	Serenade, D, pf, fl/vn (arr. of Serenade op.25: see 'Chamber Music for Wind alone and with Strings')	1803	Leipzig, 1803	arr. approved and corrected by Beethoven but largely the work of someone else	HS ix		
op.42	Notturno, D, pf, va (arr. of Serenade op.8: see 'Chamber Music for Strings')	1803	Leipzig, 1804	arr. approved and corrected by Beethoven but largely the work of someone else	HS ix		
op.69	Cello Sonata, A	1807–8	Leipzig, 1809	Baron Ignaz von Gleichenstein	xiii/107	v/3	45
op.70	Two Piano Trios, D, E♭	1808	Leipzig, 1809	Countess Marie Erdődy	xi/82–3		45
op.97	Piano Trio 'Archduke', B♭	1810–11	Vienna and London, 1816	Archduke Rudolph	xi/84		52, 62, 118, 124

op.96	Violin Sonata, G	1812, probably rev. 1815	written for Pierre Rode, ded. Archduke Rudolph	Vienna and London, 1816	xii/101	v/2	54, 118, 122
WoO 39	Allegretto, Bb, pf trio	1812	Maximiliane Brentano	Frankfurt am Main, 1830	xi/85	iv/3	55
op.102	Two Cello Sonatas, C, D	1815	Countess Erdődy	Bonn, 1817	xiii/108–9	v/3	68, 122, 126, 128
op.105	Six National Airs with Variations, pf, fl/vn	c1818		London, Edinburgh and Vienna, 1819	xiv/113–14		145
op.107	Ten National Airs with Variations, pf, fl/vn	c1818		London and Edinburgh, 1819 [nos.2, 6, 7]; Bonn and Cologne, 1820 [complete]	xiv/115–19		145

PIANO SONATAS

WoO 47	Three Sonatas ('Kurfürstensonaten'), Eb, f, D	?1783	Archbishop Maximilian Friedrich, Elector of Cologne	Speyer, 1783	xvi/156–8		4
WoO 50	Sonata, F	before 1793	Franz Gerhard Wegeler; facs. pubd in L. Schmidt: *Beethoven-Briefe* (Berlin, 1909)	Munich and Duisburg, 1950	HS ix		
op.2/1	Sonata no.1, f	1793–5	Joseph Haydn; 2nd movt uses material from Pf Qt WoO 36 no.3	Vienna, 1796	xvi/124	vii/2	20, 95
op.2/2	Sonata no.2, A	1794–5	Joseph Haydn	Vienna, 1796	xvi/125	vii/2	20, 23, 96
op.2/3	Sonata no.3, C	1794–5	Joseph Haydn; 1st movt uses material from Pf Qt WoO 36 no.3	Vienna, 1796	xvi/126	vii/2	20, 95
op.49/1	Sonata no.19, g	?1797		Vienna, 1805	xvi/142	vii/3	23, 98
op.49/2	Sonata no.20, G	1795–6		Vienna, 1805	xvi/143	vii/3	23, 98
op.7	Sonata no.4, Eb	1796–7	Countess Barbara von Keglevics	Vienna, 1797	xvi/127	vii/2	23, 98
op.10/1	Sonata no.5, c	?1795–7	Countess Anna Margarete von Browne	Vienna, 1798	xvi/128	vii/2	23, 98
op.10/2	Sonata no.6, F	1796–7	Countess von Browne	Vienna, 1798	xvi/129	vii/2	23, 98
op.10/3	Sonata no.7, D	1797–8	Countess von Browne	Vienna, 1798	xvi/130	vii/2	23, 98, 99
WoO 51	Sonata, C, frag.	completed ?1797–8	Eleonore von Breuning; 1st edn. completed by Ferdinand Ries	Frankfurt am Main, 1830	xvi/159		

No.	Title, Key	Composition	Publication	Dedication, Remarks	GA	NA	
op.13	Sonata no.8 'Pathétique', c	completed ?1797-8	Vienna, 1799	Prince Lichnowsky	xvi/131	vii/2	28, 98, 100 103, 151
op.14/1	Sonata no.9, E	1798	Vienna, 1799	Baroness Josefine von Braun	xvi/132	vii/2	
op.14/2	Sonata no.10, G	?1799	Vienna, 1799	Baroness von Braun	xvi/133	vii/2	28, 29
op.22	Sonata no.11, Bb	1800	Leipzig, 1802	Count von Browne	xvi/134	vii/2	102
op.26	Sonata no.12, Ab	1800-01	Vienna, 1802	Prince Lichnowsky	xvi/135	vii/2	
op.27/1	Sonata no.13 'quasi una fantasia', Eb	1800-01	Vienna, 1802	Princess Josephine von Liechtenstein	xvi/136	vii/3	
op.27/2	Sonata no.14, 'quasi una fantasia' ('Moonlight'), c#	1801	Vienna, 1802	Countess Giulietta Guicciardi	xvi/137	vii/3	32, 103, 151
op.28	Sonata no.15 ('Pastoral'), D	1801	Vienna, 1802	Joseph von Sonnenfels	xvi/138	vii/3	102
op.31/1	Sonata no.16, G	1802	Zurich, 1803		xvi/139	vii/3	34, 104, 113
op.31/2	Sonata no.17, d	1802	Zurich, 1803		xvi/140	vii/3	34, 98, 103, 104
op.31/3	Sonata no.18, Eb	1802	Zurich and London, 1804		xvi/141	vii/3	102, 103, 114
op.53	Sonata no.21 'Waldstein', C	1803-4	Vienna, 1805	Count Ferdinand von Waldstein	xvi/144	vii/3	37, 113
op.54	Sonata no.22, F	1804	Vienna, 1806		xvi/145	vii/3	114
op.57	Sonata no.23 ('Appassionata'), f	1804-5	Vienna, 1807	Count Franz von Brunsvik	xvi/146	vii/3	37, 43, 114, 119, 124, 151
op.78	Sonata no.24, F#	1809	Leipzig and London, 1810	Countess Therese von Brunsvik	xvi/147		50, 118
op.79	Sonata no.25, G	1809	Leipzig and London, 1810		xvi/148		50
op.81a	Sonata no.26 'Das Lebewohl, Abwesenheit und Wiedersehn', Eb	1809-10	Leipzig and London, 1811	Archduke Rudolph; Fr. sub-title 'Les adieux, l'absence et le retour'	xvi/149		49, 50, 118
op.90	Sonata no.27, e	1814	Vienna, 1815	Count Moritz Lichnowsky	xvi/150		61, 122, 126
op.101	Sonata no.28, A	1816	Vienna, 1817	Baroness Dorothea Ertmann	xvi/151		69, 122, 126, 127, 128
op.106	Sonata no.29 'Hammerklavier', Bb	1817-18	Vienna and London, 1819	Archduke Rudolph	xvi/152		72, 91, 111, 124, 126, 127, 128, 129, 133

op.109	Sonata no.30, E	1820	Berlin, 1821	Maximiliane Brentano	xvi/153		56, 74, 124, 127, 128, 134
op.110	Sonata no.31, Ab	1821–2	Paris, Berlin and Vienna, 1822; London, 1823		xvi/154		74, 126, 128, 129
op.111	Sonata no.32, c	1821–2	Paris, Berlin, Vienna and London, 1823	Archduke Rudolph; London edn. ded. Antonie Brentano	xvi/155		74, 78, 98, 124, 126, 129, 137

PIANO VARIATIONS

WoO 63	Nine Variations on a March by Dresser, c	1782	Mannheim, 1782/3	Countess Felice von Wolf-Metternich	xvii/166	vii/5	4
WoO 65	Twenty-four Variations on Righini's Arietta 'Venni amore', D	1790–9¹	?Mainz, 1791; Vienna, 1802	Countess Maria Anna Hortensia von Hatzfeld; no copy of 1791 edn. survives	xvii/178	vii/5	7, 93, 124
WoO 66	Thirteen Variations on the Arietta 'Es war einmal ein alter Mann' from Dittersdorf's Das rote Käppchen, A	1792	Bonn, 1793		xvii/175	vii/5	93
WoO 64	Six Variations on a Swiss Song, F, harp/pf	before 1793	Bonn, ?1798		xvii/177	vii/5	
WoO 68	Twelve Variations on the 'Menuet à la Viganò' from Haibel's Le nozze disturbate. C	1795	Vienna, 1796		xvii/169	vii/5	
WoO 69	Nine Variations on the Aria 'Quant' è più bello' from Paisiello's La molinara, A	1795	Vienna, 1795	Prince Lichnowsky	xvii/167	vii/5	
WoO 70	Six Variations on the Duet 'Nel cor più non mi sento' from La molinara, G	1795	Vienna, 1796		xvii/168	vii/5	
WoO 72	Eight Variations on the Romance 'Un fièvre brûlante' from Grétry's Richard Coeur de Lion, C	?1795	Vienna, 1798		xvii/171	vii/5	

No.	Title, Key	Composition	Publication	Dedication, Remarks	GA	NA	
WoO 71	Twelve Variations on a Russian Dance from Wranitzky's Das Waldmädchen, A	1796–7	Vienna, 1797	Countess von Browne	xvii/170	vii/5	
WoO 73	Ten Variations on the Duet 'La stessa, la stessissima' from Salieri's Falstaff, B♭	1799	Vienna, 1799	Countess von Keglevics	xvii/172	vii/5	
WoO 76	Six Variations on the Trio 'Tändeln und Scherzen' from Süssmayr's Soliman II, F	1799	Vienna, 1799	Countess von Browne	xvii/174	vii/5	
WoO 75	Seven Variations on the Quartet 'Kind, willst du ruhig schlafen' from Winter's Das unterbrochene Opferfest, F	1799	Vienna, 1799		xvii/173	vii/5	
WoO 77	Six Variations on an Original Theme, G	1800	Vienna, 1800		xvii/176	vii/5	96
op.34	Six Variations on an Original Theme, F	1802	Leipzig, 1803	Princess Odescalchi	xvii/162	vii/5	103
op.35	Fifteen Variations and a Fugue on an Original Theme, E♭ ('Eroica Variations')	1802	Leipzig, 1803	Count Moritz Lichnowsky; theme also used in the ballet Die Geschöpfe des Prometheus op.43, 'Eroica' Sym. op.55 and the Contredanse WoO 14 no.7	xvii/163	vii/5	28, 103
WoO 78	Seven Variations on 'God Save the King', C	1802/3	Vienna, 1804		xvii/179	vii/5	
WoO 79	Five Variations on 'Rule Britannia', D	1803	Vienna, 1804		xvii/180	vii/5	
WoO 80	Thirty-two Variations on an Original Theme, c	1806	Vienna, 1807		xvii/181	vii/5	98, 124

	Title	Composed	First published	Remarks	GA	HS/NGA	pages
op.76	Six Variations on an Original Theme, D	1809	Leipzig and London, 1810	Franz Oliva; theme used later for the Turkish March in Die Ruinen von Athen op.113	xvii/164	vii/5	
op.120	Thirty-three Variations on a Waltz by Diabelli, C	1819; 1822–3	Vienna, 1823	Antonie Brentano	xvii/165	vii/5	56, 72, 75, 78, 124, 126, 129
SHORTER PIANO PIECES							
WoO 48	Rondo, C	1783	H. P. Bossler: *Blumenlese für Klavierliebhaber*, ii (Speyer, 1783)			HS ix	
WoO 49	Rondo, A	?1783	H. P. Bossler: *Neue Blumenlese für Klavierliebhaber*, ii (Speyer, 1784)		xviii/196		96
op.39	Two Preludes through all Twelve Major Keys, pf/org	?1789	Leipzig, 1803		xviii/184		
WoO 81	Allemande, A	1793	GA		xxv/307		
op.129	Rondo a capriccio, G ('Rage over a Lost Penny')	1795	Vienna, 1828	autograph completed by unknown ed., 1828	xviii/191		
Hess 64	Fugue, C	1795	*MT*, xcvi (1955)			HS ix	
WoO 52	Presto, c	?1795	GA	probably orig. intended for Sonata op.10/1	xxv/297/1		
WoO 53	Allegretto, c	1796–7	GA	probably orig. intended for Sonata op.10/1	xxv/299		
Hess 69	Allegretto, c	1796/7	HS	probably orig. intended for Sonata op.10/1		HS ix	
op.51/1	Rondo, C	?1796–7	Vienna, 1797		xviii/185		
op.51/2	Rondo, G	?1798	Vienna, 1802	Countess Henriette Lichnowsky	xviii/186		
op.33	Seven Bagatelles, E♭, C, F, A, C, D, A♭	1801–2	Vienna and London, 1803		xviii/183		34, 79
WoO 54	? Bagatelle 'Lustig–Traurig', C	?1802	GA		xxv/300		
WoO 57	Andante, F ('Andante favori')	1803	Vienna, 1805	orig. slow movt of Sonata op.53	xviii/192		
WoO 56	Allegretto, C	1803	GA		xxv/297/2		
WoO 55	Prelude, f	before 1805	Vienna, 1805		xviii/195		

No.	Title, Key	Composition	Publication	Dedication, Remarks	GA	NA	
WoO 82	Minuet, E♭	before 1805	Vienna, 1805		xviii/193		50, 96
op.77	Fantasie, g/B♭	1809	Leipzig and London, 1810	Count Franz von Brunsvik	xviii/187		
WoO 59	Bagatelle 'Für Elise', a	1808, 1810	L. Nohl: *Neue Briefe Beethovens* (Stuttgart, 1867)	lost autograph possibly inscribed 'Für Therese' i.e. Therese Malfatti	xxv/298		
op.89	Polonaise, C	1814	Vienna, 1815	Empress Elisabeth Alexiewna of Russia	xviii/188		62
WoO 60	Bagatelle, B♭	1818	*Berliner allgemeine musikalische Zeitung*, i (1824)		xxv/301		
Hess 65	'Concert Finale', C	1820–21	in F. Starke, ed.: *Wiener Piano-Forte-Schule*, iii (Vienna, 1821)	arr. of coda to finale of Pf Conc. op.37	HS ix		
WoO 61	Allegretto, b	1821	Robitschek: *Deutscher Kunst- und Musikzeitung* (15 March 1893)	Ferdinand Piringer	HS ix		
op.119	Eleven Bagatelles, g, C, D, A, c, G, C, C, a, A, B♭	completed 1820–22	nos.7–11 in F. Starke, ed.: *Wiener Piano-Forte-Schule*, iii (Vienna, 1821); all 11, London, 1823	nos.2 and 4 sketched 1794–5; others also sketched before 1820	xviii/189		74, 79, 125
op.126	Six Bagatelles, G, g, E♭, b, G, E♭	1824	Mainz, 1825		xviii/190		79
WoO 84	Waltz, E♭	1824	Vienna, 1824	Friedrich Demmer (publisher's ded.)	xxv/303		
WoO 61a	Allegretto quasi andante, g	1825	*NZM*, cxvii (1956)	Sarah Burney Payne	HS ix		
WoO 85	Waltz, D	1825	Vienna, 1825	Duchess Sophie of Austria (publisher's ded.)	xxv/304		
WoO 86	Ecossaise, E♭	1825	Vienna, 1825	Duchess Sophie (publisher's ded.)	xxv/305		

PIANO FOUR HANDS

				Remarks		GA	NA	
WoO 67	Eight Variations on a Theme by Count Waldstein, C	?1792	Bonn, 1794			xv/122	vii/1	93
op.6	Sonata, D	1796–7	Vienna, 1797			xv/120	vii/1	
WoO 74	Six Variations on Beethoven's 'Ich denke dein', D	1799, 1803	Vienna, 1805	Countess Therese von Brunsvik and Josephine Deym (née Brunsvik); variations nos.1, 2, 5, 6 written in 1799, nos.3–4 in 1803; melody of theme (to a text by Goethe) different from that of Andenken, WoO 136 ('Ich denke dein', text by Matthisson)		xv/123	vii/1	26
op.45	Three Marches, C, Eb, D	?1803	Vienna, 1804	Princess Maria Esterházy		xv/121	vii/1	
op.134	arr. of Grosse Fuge op.133 (see 'Chamber Music for Strings')	1826	Vienna, 1827	Archduke Rudolph		HS viii	vii/1	

COLLECTIONS OF DANCE
(original scoring sometimes in doubt because versions for piano and for string trio may have been transcriptions)

No.	Title. Original scoring	Composition	Publication	Remarks	GA	NA
WoO 7	Twelve Minuets, orch	1795	for pf: Vienna, 1795; p: Vienna, ?1798; for 2 vn, b: Vienna, 1802; s: GA as WoO 7	edn. of parts sold as MS copies	ii/16 [orch]; HS viii [pf]	ii/3
WoO 8	Twelve German Dances, orch	1795	Mainz, 1933	edn. of parts sold as MS copies	ii/17 [orch]; HS viii [pf]	ii/3
WoO 9	Six Minuets, 2 vn, b	?1795	for pf: Vienna, 1796		HS vi	
WoO 10	Six Minuets	1796	Vienna, 1814		xviii/194 [pf]	ii/3
WoO 42	Six German Dances, vn, pf	?1798	for pf: Vienna, 1799		xxv/308	
WoO 11	Seven Ländler		for pf: Vienna, Prague and Leipzig, 1929	survives in pf version only; some dances sketched before 1800	xviii/198	
WoO 13	Twelve German Dances, orch				HS viii [pf]	

No.	Title, Original scoring	Composition	Publication	Remarks	GA	NA
WoO 14	Twelve Contredanses, orch	completed 1802	for pf, for 2 vn, b, and p: all Vienna, 1802; s: GA	nos.3, 4, 6 written in 1795; nos.2, 7, 9, 10 in 1801–2; others before 1802; 1st edn. for pf lacking 6 dances	ii/17a [orch]; HS viii [pf, lacking nos. 3, 6, 11]	ii/3
WoO 15	Six Ländler, 2 vn, b	1801–2	p. and for pf: Vienna, 1802; s: GA		xxv/291	ii/3
WoO 83	Six Ecossaises, pf/?orch		for pf: GA		xxv/302 [pf]	
WoO 17	Eleven Dances ('Mödlinger Tänze'), 2 fl, 2 cl, 2 hn, bn, 2 vn, b	1819	Leipzig, 1907	may be half of WoO 16, see 'Works of doubtful Authenticity' possibly spurious	HS vii	

OPERAS

No.	Title, Genre, Libretto	Composition, Production	Publication	Remarks	GA	NA
WoO 91	Two arias for Umlauf's Singspiel Die schöne Schusterin: O welch ein Leben, T solo, Soll ein Schuh nicht drükken, S solo	?1795–6	s: GA	melody of 1st aria also used for Maigesang op.52/4 (see 'Songs')	xxv/270	
Hess 115	Vestas Feuer (opera, E. Schikaneder), frag.	1803	s: Wiesbaden, 1953		HS xiii	37, 38
op.72	Fidelio oder Die eheliche Liebe ('Leonore') (opera, J. Sonnleithner, after J. N. Bouilly;	1st version (with Leonore ov. no.2), 1804–5; Theater an der Wien, Vienna, 20 Nov 1805	vs: Leipzig, 1905; s: HS	private edn. of full score, Leipzig, 1908–10; see also Table 1	HS ii, xi–xiii	40ff, 42, 105, 109, 116
	Léonore ou L'amour conjugal); (for Leonore ov. no.1, see 'Orchestral')	2nd version (with Leonore ov. no.3), 1805–6; Theater an der Wien, 29 March 1806	vs: Leipzig, 1810; s: HS	3 nos. pubd separately, Vienna, 1807	HS xi–xiii	115
		final version (with 'Fidelio' ov.), 1814; Kärntnertor-Theater, Vienna, 23 May 1814	vs: Vienna, 1814; s: Paris, 1826 (Fr.), Bonn, 1847 (Ger.)		xx/206	58f

No.	Title, Text	Composition, Production	Publication, Dedication, Remarks	GA	NA
WoO 94	Germania, finale of Die gute Nachricht (Singspiel, G. F. Treitschke), B solo, chorus	1814, Kärntnertor-Theater, 11 April 1814	vs: Vienna, 1814; s: GA · Die gute Nachricht is a pasticcio	xx/207d	
WoO 97	Es ist vollbracht, finale of Die Ehrenpforten (Singspiel, Treitschke)	1815; Kärntnertor-Theater, 15 July 1815	vs: Vienna, 1815; s: GA · Die Ehrenpforten is a pasticcio	xx/207c	

INCIDENTAL MUSIC

No.	Title, Text	Composition, Production	Publication, Dedication, Remarks	GA	NA
op.62	Overture to Collin's Coriolan (see 'Orchestral')				
op.84	Egmont (Goethe): Ov., 1 Die Trommel gerührt, song, 2–3 Entr'actes I–II, 4 Freudvoll und leidvoll, song, 5–6 Entr'actes III–IV, 7 Clärchen's Death, 8 Melodrama, 9 Siegessymphonie	1809–10; 15 June 1810	p: Leipzig, 1810 [ov.], Leipzig, 1812 [remainder]; vs: Leipzig, 1812 [without ov.]; s: Leipzig, 1831	ii/12, iii/27; HS v (no.4)	50, 116
op.113	Die Ruinen von Athen (A. von Kotzebue): Ov., 1 Tochter des mächtigen Zeus, chorus, 2 Ohne Verschulden, duet, 3 Du hast in deines Ärmels Falten, chorus of dervishes, 4 Turkish March, 5 Offstage music, 6 Schmükt die Altäre, march and chorus, 7 Mit reger Freude, recit and aria, 8 Heil unserm König, heil!, chorus	1811; 10 Feb 1812	no.4 for pf 4 hands: Vienna, 1822–3; s: Vienna, 1823 [ov. only], Vienna, 1846 [complete, ded. (by publisher) Kaiser Friedrich Wilhelm IV of Prussia]	xx/207, iii/28	52
op.117	König Stephan (Kotzebue): Ov., 1 Ruhend von seinen Taten, chorus, 2 Auf dunklem Irrweg, chorus, 3 Siegesmarsch, 4 Wo die Unschuld Blumen streute, chorus, 5 Melodrama, 6 Eine neue strahlende Sonne, chorus, 7 Melodrama, 8 Heil unserm Königel, geistliche Marsch, chorus and melodrama, 9 Heil unserm Enkeln!, chorus	1811; 10 Feb 1812	ov. and no.3 for pf 4 hands: Vienna, 1822–3; s: Vienna, 1826 [ov. only], GA [complete]; see also WoO 98	xx/207b, iii/23	52
WoO 2a	Triumphal March, C, for Tarpeja (C. Kuffner)	1813; 26 March 1813	for pf: Vienna, 1813; p: Vienna, 1840; s: GA	ii/14	

175

No.	Title, Text	Composition, Production	Publication, Dedication, Remarks	GA	NA	
WoO 2b	'Introduction to Act 2', ? for Tarpeja	1813; 26 March 1813	s: Mainz, 1938; no direct evidence for associating this with Tarpeja	HS iv		
WoO 96	Leonore Prohaska (F. Duncker): 1 Wir bauen und sterben, chorus, 2 Es blüht eine Blume, romance, 3 Melodrama, 4 Funeral March	1815	s: GA; no.4 is arr. of Funeral March from Piano Sonata op.26	xxv/272		
op.124	Overture, C, to Die Weihe des Hauses (C. Meisl)	1822; Joseph-stadt-Theater, 3 Oct 1822	s: Mainz, 1825; ded. Prince Golitsin	iii/24	ii/1	75, 78, 79, 82, 126
WoO 98	Wo sich die Pulse, chorus for Die Weihe des Hauses	as op.124	s: GA; Die Weihe des Hauses is adaptation of Die Ruinen von Athen and incorporates new or revised texts for op.113 no.1 (pubd in HS xiii), no.6 (pubd as op.114, Vienna, 1826), no.7 and no.8	xxv/266		75

BALLETS

No.	Title, Choreographer	Composition, Production	Publication, Dedication, Remarks	GA	NA	
WoO 1	Ritterballett (Count Waldstein, ? collab. Habich): 1 March, 2 German Song, 3 Hunting-song, 4 Love-song ('Romance'), 5 War Song, 6 Drinking-song, 7 German Dance, 8 Coda	1790–91; Bonn, 6 March 1791	for pf: Leipzig and Winterthur, 1872; s: GA	xxv/286; HS viii [pf]	ii/1	7
op.43	Die Geschöpfe des Prometheus (S. Viganò), ov., introduction and 16 numbers	1800–01; Burgtheater, Vienna, 28 March 1801	for pf: Vienna, 1801; p: Leipzig, 1804 [ov. only]; s: GA; ded. Princess Christiane von Lichnowsky	ii/11; HX viii [pf]	ii/1	28, 103, 104, 109

CHORAL WORKS WITH ORCHESTRA

No.	Title. Scoring	Composition	Publication	Dedication. Remarks	GA	NA
WoO 87	Cantata on the death of the Emperor Joseph II (S. A. Averdonk), S, A, T, B, 4vv; orch	1790	GA		xxv/264	6, 10
WoO 88	Cantata on the accession of Emperor Leopold II (Averdonk), S, A, T, B, 4vv, orch	1790	GA		xxv/265	7, 10
op.85	Christus am Oelberge [The Mount of Olives] (F. X. Huber), oratorio, S, T, B, 4vv, orch	1803, rev. 1804	Leipzig, 1811		xix/205	35, 36, 105, 115
op.86	Mass, C, S, A, T, B, 4vv	1807	Leipzig, 1812	Prince Ferdinand Kinsky	xix/204	44, 46, 115
op.80	Fantasia, c, pf, chorus, orch (see 'Solo Instruments and Orchestra')					
WoO 95	Chor auf die verbündeten Fürsten 'Ihr weisen Gründer' (C. Bernard), 4vv, orch	1814	GA	for Congress of Vienna	xxv/267	61, 91
op.136	Der glorreiche Augenblick (A. Weissenbach), cantata, 2 S, T, B, 4vv, orch	1814	Vienna, 183?	for Congress of Vienna; also pubd with new text by F. Rochlitz as Preis der Tonkunst (Vienna, 1837)	xxi/208	61, 91
op.112	Meeresstille und glückliche Fahrt (Goethe), ?cantata, 4vv, orch	1814-15	Vienna, 182?	Johann Wolfgang von Goethe	xxi/209	
op.123	Mass, D ('Missa solemnis'), S, A, T, B, 4vv, orch, org	1819-23	Mainz, 1827	Archduke Rudolph; orig. intended for Rudolph's installation as cardinal, 9 March 1820	xxi/203	72, 129ff, 137, 143
—	Opferlied 'Die Flamme lodert' (F. von Matthisson), S, A, T, 4vv, 2 cl, hn, va, vc	1822	GA		xxv/268	
op.121b	Opferlied, 2nd version, S, 4vv, orch	1823-4	Mainz, 1825	version with pf acc. pubd in HS v; see also 'Songs', WoO 126	xxii/212	79
op.122	Bundeslied 'In allen guten Stunden' (Goethe), S, A, 3vv, 2 cl, 2 hn. 2 bn	1823-4	Mainz, 1825	version with pf acc. pubd in HS v	xxii/213	
op.125	Symphony no.9, d (see 'Orchestral')					

OTHER CHORAL WORKS

No.	Title, Scoring	Composition	Publication, Dedication, Remarks	GA	NA
WoO 102	Abschiedsgesang 'Die Stunde schlägt' (J. von Seyfried), 2 T, B	1814	GA; Leopold Weiss; at the request of Mathias Tuscher	xxv/273	
WoO 103	Cantata campestre 'Un lieto brindisi' (Abbate Clemente Bondi), S, 2 T, B, pf	1814	Jb der Literarischen Vereinigung Winterthur 1945; Giovanni Malfatti; ? at the request of Andreas Bertolini	HS v	
WoO 104	Gesang der Mönche 'Rasch tritt der Tod', from Wilhelm Tell (Schiller), 2 T, B	1817	NZM, vi (1839); in memory of Franz Sales Kandler and Wenzel Krumpholz	xxiii/255	
WoO 105	Hochzeitslied 'Auf Freunde, singt dem Gott: der Ehen' (A. J. Stein), 2 versions: C major, T, unison male vv, pf	1819	Der Bär 1927; Anna Giannatasio del Rio	HS v	
	A major, male solo v, 4vv, pf	?1819	London, 1858; Anna Giannatasio del Rio	HS v	
WoO 106	Birthday Cantata for Prince Lobkowitz 'Es lebe unser theurer Fürst' (Beethoven), S, 4vv, pf	1823	L. Nohl: Neue Briefe Beethovens (Stuttgart, 1867); Prince Lobkowitz	xxv/274	

SOLO VOICES AND ORCHESTRA

No.	Title, Scoring	Composition	Publication, Dedication, Remarks	GA	NA
WoO 89	Prüfung des Küssens 'Meine weise Mutter spricht', aria, B solo	c1790–92	GA	xxv/269/1	93
WoO 90	Mit Mädeln sich vertragen, aria, from Claudine von Villa Bella (Goethe), B solo	c1790–92	GA	xxv/269/2	93
WoO 92	Primo amore, scena and aria, S solo	c1790–92	GA	xxv/271	8, 93
WoO 91	Two arias for Die schöne Schusterin (see 'Operas')				
op.65	Ah! perfido, scena and aria, recit from Achille in Sciro (Metastasio), S solo	1795–6	p. vs: Leipzig, 1805; s: GA; Countess Josephine de Clary (ded. in MS, not in 1st edn.)	xxii/210	46, 99
WoO 92a	No, non turbarti, scena and aria, from La tempesta (Metastasio), S solo	1801–2	Wiesbaden, 1949	HS ii	15
WoO 93	Ne' giorni tuoi felici, duet, from Olimpiade (Metastasio), S, T	1802–3	Leipzig, 1939	HS ii, xiv	15
op.116	Tremate, empi, tremate (Bettoni), S, T, B	1801–2; 1814	p. vs: Vienna, 1826; s: GA	xxii/211	15
op.118	Elegischer Gesang 'Sanft wie du lebtest', S, A, T, B, str qt/pf	1814	Vienna, 1826 [with pf acc., separate parts for str qt]; Baron Johann von Pasqualati	xxii/214	

No.	Title, Author of text	Text incipit	Composition	Publication, Dedication, Remarks	GA	NA
WoO 107	Schilderung eines Mädchens	Schildern, willst du Freund, soll ich dir Elisen?	?1783	H. P. Bossler: *Blumenlese für Klavierliebhaber*, ii (Speyer, 1783)	xxiii/228	
WoO 108	An einen Säugling (? J. von Döhring)	Noch weisst du nicht wess Kind du bist	?1784	H. P. Bossler: *Neue Blumenlese für Klavierliebhaber*, ii (Speyer, 1784)	xxiii/229	
WoO 113	Klage (L. Hölty)	Dein Silber schien durch Eichengrün	?1790	GA	xxv/283	
WoO 109	Trinklied	Erhebt das Glas	?1790	GA	xxv/282	
WoO 111	Punschlied, with unison vv	Wer nicht, wenn warm von Hand zu Hand	c1790–92	L. Schiedermair: *Der junge Beethoven* (Leipzig, 1925)	HS v	
Hess 151	Traute Henriette		c1790–92	*ÖMz*, iv (1949)	HS v	
WoO 112	An Laura (F. von Matthisson)	Freud' umblühe dich auf allen Wegen	?1792	G. Kinsky: *Musik-historisches Museum von Wilhelm Heyer in Cöln: Katalog*, iv (Cologne, 1916); pf arr. pubd (as Bagatelle, op.119/12) Vienna, ?1826	HS v	
WoO 114	Selbstgespräch (J. W. L. Gleim)	Ich, der mit flatterndem Sinn	?1792	GA	xxv/275	
WoO 115	An Minna	Nur bei dir, an deinem Herzen	?1792	GA	xxv/280	
WoO 110	Elegie auf den Tod eines Pudels	Stirb immerhin, es welken ja so viele der Freuden	? before 1793	?GA; may have been pubd by the 1830s	xxv/284	
WoO 116	Que le temps me dure (J.-J. Rousseau) 1st version 2nd version		1793 1793	*Die Musik*, i (1901–2) *ZfM*, cii (1935)	HS v HS v	
Hess 129 Hess 130						
WoO 117	Der freie Mann (G. C. Pfeffel), with unison vv	Wer ist ein freier Mann?	1792, rev. 1794	Bonn, 1808	xxiii/232; HS v	
WoO 119	O care selve (Metastasio), with unison vv		1794	GA	xxv/279	
WoO 126	Opferlied (Matthisson)	Die Flamme lodert	1794, rev. 1801–2	Bonn, 1808; see also 'Choral Works with Orchestra'	xxiii/233; HS v	

179

No.	Title, Author of text	Text incipit	Composition	Publication, Dedication, Remarks	GA	NA
WoO 118	Two songs (G. A. Bürger): 1 Seufzer eines Ungeliebten, 2 Gegenliebe	1 Hast du nicht Liebe zugemessen, 2 Wüsst ich, dass du mich lieb	1794–5	Vienna, 1837	xxiii/253	
op.46	Adelaide (Matthisson)	Einsam wandelt dein Freund im Frühlings Garten	1794–5	Vienna, 1797; Friedrich von Matthisson	xxiii/216	23, 99
WoO 123	Zärtliche Liebe (K. F Herrosee)	Ich liebe dich	?1795	Vienna, 1803	xxiii/249	
WoO 124	La parenza (Metastasio)	Ecco quel fiero istante!	?1795–6	Vienna, 1803	xxiii/251	
WoO 121	Abschiedsgesang an Wiens Bürger (Friedelberg)	Keine Klage soll erschallen	1796	Vienna, 1796; Obrist Wachtmeister von Kövesdy	xxiii/230	
WoO 122	Kriegslied der Österreicher (Friedelberg), with unison vv	Ein grosses deutsches Volk sind wir	1797	Vienna, 1797	xxiii/231	
WoO 127	Neue Liebe, neues Leben (Goethe)	Herz, mein Herz, was soll das geben?	1798/9	Bonn, 1808; same text also set as op.75/2	HS v	
WoO 125	La tiranna (? trans. W. Wennington)	Ah grief to think	1798–9	London, 1799; ded. (by Wennington) Mrs Tschoffen	HS v	25
WoO 128	Plaisir d'aimer		1798–9	Die Musik, i (1901–2)	HS v	
WoO 74	Ich denke dein (see 'Piano Four Hands')					
WoO 120	Man strebt die Flamme zu verheblen		c1800–02	GA	xxv/278	
op.48	Six Songs (C. F. Gellert): 1 Bitten	Gott, deine Güte reicht so weit	before March 1802	Vienna, 1803; Count von Browne; no.3 sketched 1798	xxiii/217; HS v [no.6]	32
	2 Die Liebe des Nächsten	So jemand spricht: ich liebe Gott				
	3 Vom Tode	Meine Lebenszeit verstreicht				
	4 Die Ehre Gottes aus der Natur	Die Himmel rühmen				
	5 Gottes Macht und Vorsehung	Gott ist mein Lied				
	6 Busslied	An dir allein, an dir hab' ich gesündigt				

WoO 129	Der Wachtelschlag (S. F. Sauter)	Ach mir schallt's dorten	1803	Vienna, 1804; Count von Browne	xxiii/234	150
op.88	Das Glück der Freundschaft	Der lebt ein Leben wonniglich	1803	Vienna, 1803	xxiii/222	
op.52	Eight Songs				xxiii/218; HS v [no.2]	92
	1 Urians Reise um die Welt (M. Claudius), with unison vv	Wenn jemand eine Reise tut	before 1793	Vienna, 1805		
	2 Feuerfarb (S. Mereau)	Ich weiss eine Farbe	1792, rev. 1793–4			
	3 Das Liedchen von der Ruhe (H. W. F. Ueltzen)	Im Arm der Liebe	1793			
	4 Maigesang (Goethe)	Wie herrlich leuchtet mir die Natur	? before 1796	theme also used in WoO 91 no.1 (see 'Operas')		
	5 Mollys Abschied (Bürger)	Lebe wohl, du Mann der Lust und Schmerzen				
	6 Die Liebe (G. E. Lessing)	Ohne Liebe lebe wer da kann	before 1793			
	7 Marmotte (Goethe)	Ich komme schon durch manche Land	?c1790–92			
	8 Das Blümchen Wunderhold (Bürger)	Es blüht ein Blümchen irgendwo				
op.32	An die Hoffnung (C. A. Tiedge)	Die du so gern in heilgen Nächten feierst	1805	Vienna, 1805; see also op.94	xxiii/215	
WoO 132	Als die Geliebte sich trennen wollte (?Hoffmann, trans. (from Fr.) S. von Breuning)	Der Hoffnung letzter Schimmer	1806	AMZ, xii (1809–10); also pubd as Empfindung bei Lydiens Untreue	xxiii/235	

150

No.	Title, Author of text	Text incipit	Composition	Publication, Dedication, Remarks	GA	NA
WoO 133	In questa tomba oscura (G. Carpani)	—	1807	Vienna, 1808; ded. (by publisher) Prince Lobkowitz	xxiii/252	
WoO 134	Sehnsucht (Goethe), 4 settings	Nur wer die Sehnsucht kennt	1807–8	Vienna, 1810; no.1 first pubd in *Prometheus*, no.3 (1808)	xxiii/250	
WoO 136	Andenken (Matthisson)	Ich denke dein	1809	Leipzig and London, 1810	xxiii/248	
WoO 137	Lied aus der Ferne (C. L. Reissig)	Als mir noch die Thräne	1809	Leipzig and London. 1810; text orig. used for WoO 138, pubd in HS v	xxiii/236	
WoO 138	Der Jüngling in der Fremde (Reissig)	Der Frühling entblühet	1809	Vienna, 1810; ded. (by Reissig) Archduke Rudolph; orig. written to text of WoO 137, pubd in HS v	xxiii/237	
WoO 139	Der Liebende (Reissig)	Welch ein wunderbares Leben	1809	Vienna and London, 1810; ded. (by Reissig) Archduke Rudolph	xxiii/238	
op.75	Six Songs			Leipzig and London, 1810; Princess Caroline Kinsky	xxiii/219	
	1 Mignon (Goethe)	Kennst du das Land	1809			
	2 Neue Liebe, neues Leben (Goethe)	Herz, mein Herz, was soll das geben?	1809	text set previously in WoO 127		
	3 Aus Goethes Faust, with unison vv	Es war einmal ein König	1809	sketched c1792		
	4 Gretels Warnung (G. A. von Halem)	Mit Liebesblick und Spiel und Sang				
	5 An den fernen Geliebten (Reissig)	Einst wohnten süsse Ruh	1809			
	6 Der Zufriedene (Reissig)	Zwar schuf das Glück hienieden	1809			

182

51

op.82	Four Ariettas and a Duet, S, T		?1809	Leipzig and London, 1811; may have been written c1801	xxiii/220; HS v [no.1]
	1 Hoffnung	Dimmi ben mio			
	2 Liebes-Klage (Metastasio)	T'intendo, sì, mio cor			
	3 L'amante impaiente (Metastasio), arietta buffa	Che fà il mio bene?			
	4 L'amante impaiente (Metastasio), arietta assai seriosa	Che fà il mio bene?			
	5 Lebens-Genuss (Metastasio), duet	Odi l'aura che dolce sospira			
op.83	Three Songs (Goethe)		1810	Leipzig, 1811; Princess Caroline Kinsky	xxiii/221: HS v [no.1]
	1 Wonne der Wehmut	Trocknet nicht			
	2 Sehnsucht	Was zieht mir das Herz so?			
	3 Mit einem gemalten Band	Kleine Blumen, kleine Blätter			
WoO 140	An die Geliebte (J. L. Stoll), 2 versions	O dass ich dir vom stillen Auge	1811; ?1814	1st version (pf/gui acc.): Augsburg, after 1825; 2nd version pubd in *Friedensblätter* (12 July 1814) GA	xxiii/243a, 243
WoO 141	Der Gesang der Nachtigal (J. G. Herder)	Höre, die Nachtigall singt	1813		xxv/277
WoO 142	Der Bardengeist (F. R. Hermann)	Dort auf dem hohen Felsen sang	1813	Erichson: *Musen-Almanach für das Jahr 1814* (Vienna, 1813–14)	xxiii/241
WoO 143	Des Kriegers Abschied (Reissig)	Ich zieh' ins Feld	1814	Vienna, 1815: ded. (by Reissig) Caroline von Bernath	xxiii/240

No.	Title, Author of text	Text Incipit	Composition	Publication, Dedication, Remarks	GA	NA
WoO 144	Merkenstein (J. B. Rupprecht)	Merkenstein! Wo ich wandle denk' ich dein	1814	*Selam: ein Almanach für Freunde des Mannigfaltigen auf das Schaltjahr 1816* (Vienna, 1815–16); see also op.100	xxv/276	
op.100	Merkenstein (J. B. Rupprecht), duet	Merkenstein! Wo ich wandle denk' ich dein	1814–15	Vienna, 1816; Count Joseph Karl von Dietrichstein; see also WoO 144	xxiii/226	
op.94	An die Hoffnung (C. A. Tiedge)	Ob ein Gott sei	?1815	Vienna, 1816; Princess Kinsky; sketched 1813	xxiii/223	
WoO 135	Die laute Klage (Herder)	Turteltaube, du klagtest so laut	?c1815	Vienna, 1837	xxiii/254	
WoO 145	Das Geheimnis (I. von Wessenberg)	Wo blüht das Blümchen	1815	*Wiener Zeitschrift für Kunst, Literatur, Theater und Mode*, i (1816)	xxiii/245	
WoO 146	Sehnsucht (Reissig)	Die stille Nacht umdunkelt	1815–16	Vienna, 1816	xxiii/239	
op.98	An die ferne Geliebte (A. Jeitteles), cycle of 6 songs	1 Auf dem Hügel sitz ich spähend 2 Wo die Berge so blau 3 Leichte Segler in den Höhen 4 Diese Wolken in den Höhen 5 Es kehret der Maien 6 Nimm sie hin denn diese Lieder	1815–16	Vienna, 1816; Prince Lobkowitz	xxiii/224	69, 122, 126, 151
op.99	Der Mann von Wort (F. A. Kleinschmid)	Du sagtest, Freund, an diesen Ort	1816	Vienna, 1816	xxiii/225	
WoO 147	Ruf vom Berge (G. F. Treitschke)	Wenn ich ein Vöglein wär	1816	*Gedichte von Friedrich Treitschke* (Vienna, 1817)	xxiii/242	122
WoO 148	So oder so (C. Lappe)	Nord oder Süd!	1817	*Wiener Zeitschrift für Kunst*, ii (1817)	xxiii/244	

No.	Incipit, No. of parts, Recipient or Occasion	Composition	Publication, Remarks	GA	NA
WoO 149	Resignation (P. von Haugwitz)	Lisch aus, lisch aus, mein Licht!	1817	Wiener Zeitschrift für Kunst, iii (1818); orig. sketches for 4vv, 1816	xxiii/246	
WoO 130	Gedenke mein	—		Vienna, 1844	xxv/281	
WoO 150	Abendlied unterm gestirnten Himmel (H. Goeble)	Wenn die Sonne nieder sinket	?1804–5, rev. 1815–20	Wiener Zeitschrift für Kunst, v (1820); Anton Braunhofer	xxiii/247	
op.128	Der Kuss (C. F. Weisse)	Ich war bei Chloen ganz allein	?1822	Mainz, 1825; sketched 1798	xxiii/227	
WoO 151	Der edle Mensch sei hülfreich und gut (Goethe)		1823	G. Lange: Musikgeschichtliches (Berlin, 1900), facs. in Allgemeine Wiener Musik-Zeitung (23 Nov 1843); written for Baroness Cäcilie von Eskeles	HS v	

CANONS AND MUSICAL JOKES

No.	Incipit, No. of parts, Recipient or Occasion	Composition	Publication, Remarks	GA	NA
WoO 159	Im Arm der Liebe, 3vv, contrapuntal study for Albrechtsberger	?1795	I. von Seyfried: Ludwig van Beethovens Studien im Generalbass (Vienna, 1832)	xxiii/256/1	
WoO 160/1	? O care selve, 3vv, contrapuntal study for Albrechtsberger	?1795	Seyfried (1832)	HS v	
WoO 160/2	Canon, 4vv, contrapuntal study for Albrechtsberger	?1795	Seyfried (1832)	HS v	
—	Canon, 3vv	1796–7	J. Kerman, ed.: Ludwig van Beethoven: Autograph Miscellany from circa 1796 to 1799 (London, 1970)	—	
Hess 276	Herr Graf, ich komme zu fragen, 3vv	?1797	HS; also sketched with different text Grove I ('Schuppanzigh, Ignaz')	HS v	
WoO 100	Schuppanzigh ist ein Lump, T, 2 B, 4vv (not canonic), for Ignaz Schuppanzigh	1801		HS v	
WoO 101	Graf, Graf, Graf, Graf, 3vv (not canonic), for Nikolaus Zmeskall von Domanovecz	1802	A. W. Thayer: Chronologisches Verzeichniss der Werke Ludwig van Beethovens (Berlin, 1865)	HS v	

No.	Incipit, No. of parts, Recipient or Occasion	Composition	Publication, Remarks	GA	NA
Hess 274	Canon, 2vv	1803	N. Fishman: *Kniga eskizov Beethoven za 1802–1803 gody* (Moscow, 1962)	HS ix	
Hess 229	Languisco e moro, 2vv	1803	Fishman (1962); also sketched as song for 1v, pf	HS xiv	
Hess 275	Canon, 2vv	1803	HS	HS ix	
WoO 162	Ta ta ta ... lieber Mälzel, 4vv	1813	see 'Works of Doubtful Authenticity' *NZM*, xi (1841), suppl	xxiii/256/3a	
WoO 163	Kurz ist der Schmerz, 3vv, for Johann Friedrich Naue	1814	GA	xxv/285/2	
WoO 164	Freundschaft ist die Quelle, 3vv	1814	Vienna, 1816	xxiii/256/16	
WoO 165	Glück zum neuen Jahr, 4vv, for Baron von Pasqualati	1815	GA; facs. in L. Spohr: *Selbstbiographie* (Kassel and Göttingen, 1860)	xxiii/256/3b	
WoO 166	Kurz ist der Schmerz, 3vv, for Louis Spohr	1815	*Wiener allgemeine musikalische Zeitung*, i (1816)	xxiii/256/5	
WoO 168/1	Lerne schweigen, puzzle canon (?3vv), for Charles Neate	1815–16	GA	xxiii/256/4	
WoO 168/2	Rede, rede, 3vv, for Neate	1815–16	*Die Jahreszeiten*, xii/3 (1853)	HS v	
WoO 169	Ich küsse Sie, puzzle canon (?2vv), for Anna Milder-Hauptmann	1816	L. Nohl: *Neue Briefe Beethovens* (Stuttgart, 1867)	HS v	
WoO 170	Ars longa, vita brevis, 2vv, for Johann Nepomuk Hummel	1816	T. von Frimmel: *Neue Beethoveniana* (Vienna, 1888)	HS v	
WoO 171	Glück fehl' dir vor allem. 4vv, for Anna Giannatasio del Rio	1817	Thayer (1865)	HS v	
WoO 173	Hol' euch der Teufel!, puzzle canon, (?2vv), for Sigmund Anton Steiner	1819	L. Nohl: *Briefe Beethovens* (Stuttgart, 1865); facs. in A. B. Marx: *Ludwig van Beethoven: Leben und Schaffen* (Berlin, 1859), ii		
WoO 174	Glaube und hoffe, 4vv (not canonic). for Maurice Schlesinger	1819	GA	xxv/285/3	
WoO 176	Glück zum neuen Jahr!, 3vv, for Countess Erdödy	1819	GA	xxiii/256/6	
WoO 179	Alles Gute! alles Schöne, 4vv, for Archduke Rudolph	1819	Nohl (1865); incl. non-canonic introduction 'Seiner kaiserlichen Hoheit'	xxiii/256/7	
Hess 300	Liebe mich, werter Weissenbach, ? for Aloys Weissenbach	1819–20	J. Schmidt-Görg, ed.: *Drei Skizzenbücher zur Missa Solemnis*, i (Bonn, 1952)	—	

WoO 175	Sankt Petrus war ein Fels; Bernardus war ein Sankt, puzzle canons (?4vv), for Carl Peters and Carl Bernard	1819–20	Thayer (1865); 2nd canon based on melody of 1st, in rhythmic augmentation	HS v
WoO 180	Hoffmann, sei ja kein Hoffmann, 2vv	1820	*Caecilia*, i (1825)	xxiii/256/8
WoO 181/1	Gedenket heute an Baden, 4vv	?1820	GA	xxv/285/4
WoO 181/2	Gehabt euch wohl, 4vv	?1820	*Festschrift Arnold Scherings* (Berlin, 1937)	HS v
WoO 181/3	Tugent ist kein leerer Name, 3vv	?1820	as WoO 181/2	HS v
WoO 182	O Tobias!, 3vv, for Tobias Haslinger	1821	*AMZ*, new ser., i (1863)	xxiii/256/9
WoO 183	Bester Herr Graf, Sie sind ein Schaf!, 4vv, for Count Moritz Lichnowsky	1823	*Mf*, vii (1954); facs. in *Musikalisch-kritisches Repertorium*, i/10 (Leipzig, 1844); inaccurate edn. in A. W. Thayer: *Ludwig van Beethovens Leben*, ed. H. Riemann, iv (Leipzig, 1907)	HS v
WoO 184	Falstafferel, lass' dich sehen!, 5vv, for Schuppanzigh	1823	*Die Musik*, ii (1902–3)	HS v
WoO 185	Edel sei der Mensch, 6vv, for Louis Schlösser	1823	*Wiener Zeitschrift für Kunst*, viii (1823) [in E major; Beethoven also wrote out version in E♭]; a canon in E♭ for 3vv on the text 'Edel hülfreich sei der Mensch' sketched in 1822	xxiii/256/10
Hess 263	Te solo adoro, 2vv, ? for Carlos Evasio Soliva	?1824	HS; similar to (? and earlier version of) WoO 186	HS v
Hess 264	Te solo adoro, 2vv, ? for Soliva	?1824	HS; similar to (? and earlier version of) WoO 186	HS v
WoO 186	Te solo adoro, 2vv, for Soliva	1824	GA	xxv/285/1
WoO 187	Schwenke dich ohne Schwänke!, 4vv, for Carl Schwencke	1824	*Caecilia*, i/7 (1825)	xxiii/256/11
WoO 188	Gott ist eine feste Burg, puzzle canon (?2vv), for Oberst von Düsterlohe	1825	F. Prelinger: *Beethovens sämtliche Briefe*, iv (Vienna, 1909); facs. in auction catalogue no.36 of Leo Liepmannssohn (Berlin, 1906)	HS v
WoO 203	Das Schöne zu dem Guten, puzzle canon (?4vv), for Ludwig Rellstab	1825	L. Rellstab: *Garten und Wald*, iv (Leipzig, 1854); WoO 202 is 2-bar non-canonic greeting on the same text	HS v
WoO 189	Doktor, sperrt das Tor dem Tod, 4vv, for Anton Braunhofer	1825	Nohl (1865)	HS v
WoO 190	Ich war hier, Doktor, puzzle canon (?2vv), for Braunhofer	1825	?HS; facs. in auction catalogue no.21 of M. Breslauer (Berlin, 1912)	HS v
WoO 35	Canon, 2vv (? for 2 vn), for Otto de Boer	1825	Nohl (1867)	HS vi

137

No.	Incipit. No. of parts. Recipient or Occasion	Composition	Publication, Remarks	GA	NA
WoO 191	Kühl, nicht lau, 3vv, for Friedrich Kuhlau	1825	Seyfried (1832)	xxiii/256/12	
WoO 192	Ars longa, vita brevis, puzzle canon (?4vv), for Sir George Smart	1825	Thayer (1865)	HS v	
WoO 194	Si non per portas, per muros, puzzle canon (?2vv), for Maurice Schlesinger	1825	Marx (1859), ii	xxiii/256/17	
WoO 204	Holz, Holz, geigt die Quartette so, lv, for Karl Holz	1825	A. W. Thayer: *Ludwig van Beethovens Leben*, ed. H. Riemann, v (Leipzig, 1908)	—	
WoO 195	Frau' dich des Lebens, 2vv, for Theodor Molt	1825	GA	xxv/285/5	
Hess 277	Esel aller Esel, 2 canonic vv, ostinato v	1826	HS	HS v	
WoO 196	Es muss sein, 4vv, for 'Hofkriegsagent' Dembscher	1826	A. W. Thayer: *Ludwig van Beethovens Leben*, ed. H. Riemann, v (Leipzig, 1908); facs. in Gassner: *Zeitschrift für Deutschlands Musikvereine und Dilettanten*, iii (Karlsruhe, 1844)	HS v	
—	Bester Magistrat, 3vv	1826	unpubd, appears in the sketchbook 'Autograph 24', *D-B*	—	
WoO 197	Da ist das Werk, 5vv, for Holz	1826	Zurich, 1949	HS v	
WoO 198	Wir irren allesamt, puzzle canon (?2vv), for Holz	1826	Nohl (1865)	HS v	
WoO 161	Ewig dein, 3vv	—	*AMZ*, new ser., i (1863); ? written c1810	xxiii/256/14	
WoO 167	Brauchle, Linke, ?4vv, ? for Johann Xaver Brauchle and Joseph Linke	—	Thayer (1865); ? written c1815	HS v	
WoO 172	Ich bitt' dich, schreib' mir die Es-Scala auf, 3vv, for Vincenz Hauschka	—	GA; ? written c1818	xiii/256/15	
WoO 177	Bester Magistrat, Ihr friert, ?2/4vv, bass v	—	D. MacArdle and L. Misch: *New Beethoven Letters* (Norman, Oklahoma, 1957); facs. in auction catalogue no.132 of K. E. Henrici (Berlin, 1928); ? written c1820	HS v	
WoO 178	Signor Abate, 3vv, for the Abbé Maximilian Stadler	—	GA	xxiii/256/13	
WoO 193	Ars longa, vita brevis, puzzle canon (?5vv)	—	facs. in auction catalogue no.120 of Henrici (Berlin, 1927)	HS v	

No.	Work	Composition	Publication, Dedication, Remarks	GA	NA
WoO 33/1	Adagio, F, mechanical clock	?c1799	*Die Musik*, i (1901–2)	HS vii	
WoO 33/2	Scherzo, G, mechanical clock	1799–1800	G. Becking: *Studien zu Beethovens Personalstil: das Scherzothema* (Leipzig, 1921)	HS vii	
WoO 33/3	Allegro, G, mechanical clock	?c1799	*Ricordiana*, iii (1957)	HS vii	
WoO 33/4	Allegro, C, ? mechanical clock	?1794	Mainz, 1940	HS vii	
WoO 33/5	Minuet, C, ? mechanical clock	?1794	Mainz, 1940	HS vii	
Hess 107	Grenadiermarsch, F, mechanical clock	?c1798	*Beethoven-Almanach der Deutschen Musikbücherei auf das Jahr 1927* (Regensburg, 1927); Prince Joseph Johann zu Schwarzenberg; consists of march by Haydn, transition section by Beethoven and transcr. of WoO 29 (see 'Chamber Music for Wind alone and with Strings')	HS vii	
WoO 58	Cadenzas to 1st movt and finale of Mozart's Pf Conc., d, k466	?1809	1st movt: *Wiener Zeitschrift für Kunst . . .* (23 Jan 1836); finale: GA	ix/70a/11–12	vii/7
—	Contrapuntal exercises prepared for Haydn and Albrechtsberger (see Hess 29–31, 233–46)	1792–5	G. Nottebohm: *Beethovens Studien* (Leipzig and Winterthur, 1873), selective transcr.	HS vi, xiv	
—	Exercises in Italian declamation prepared for Salieri (see WoO 99; Hess 208–232)	c1801–2	Nottebohm (1873) [selective]; HS i [complete]; WoO 92a and WoO 93 may have been the culminating studies (see 'Solo Voices and Orchestra')	HS i	
—	Various dances, kbd exercises, entered among sketches for larger works but probably not intended for publication (? incl. WoO 81; Hess 58–61, 67–8, 70–74, 312–34)	mostly 1790–98	transcr. selectively in writings of Nottebohm (see Bibliography); many pubd in Kerman, ed. (1970)	(HS ix)	
—	Various musical greetings, in letters and diaries etc (see WoO 205; Hess 278–95); see also 'Canons and Musical Jokes'				
WoO 200	Theme for variations by the Archduke Rudolph, with text 'O Hoffnung'	1818	Vienna, 1819 (Rudolph's set of variations)		

no.

WoO 27 Three Duets, cl, bn, C, F, Bb (Paris, ?c1810–15); probably spurious; GA viii/64

— Flute Sonata, Bb, ?c1790–92 (Leipzig, 1906), listed as Anhang 4 in G. Kinsky and H. Halm: *Das Werk Beethovens* (Munich and Duisburg, 1955), MS copy found among Beethoven's papers after his death, but authenticity not certain; HS ix

WoO 12 Twelve Minuets, orch, 1799 (for pf: Paris, 1903, s: Paris, 1906); probably by Beethoven's brother Carl; HS iv

WoO 16 Twelve Ecossaises, orch, advertised Vienna, 1807; no copy survives; these Ecossaises and 12 waltzes are foreign arrangements of movements from Beethoven works

WoO 17 Eleven Dances, 2 fl, 2 cl, 2 hn, bn, 2 vn, b, see 'Collections of Dances'

WoO 162 Ta ta ta . . . lieber Mälzel, 4vv, for Johann Nepomuk Maelzel; *Musikalisch-kritisches Repertorium aller neuen Erscheinungen im Gebiete der Tonkunst*, ed. H. Hirschbach, i/2 (Leipzig, 1844); GA xxiii/256/2; ?forgery by Schindler, see Howell (1979)

FOLKSONG ARRANGEMENTS
(with pf trio acc. unless otherwise stated)

Beethoven appears to have begun arranging folksongs for the Scottish publisher George Thomson in late 1809 or early 1810. He continued to do so at intervals until at least 1816. It was his own idea to extend the scope of the project to include songs not of British origin; most of these were never published by Thomson. In the lists below, the songs are grouped as they appear in GA xxiv and many later publications.

no.

WoO 152 Twenty-five Irish songs (London and Edinburgh, 1814); GA xxiv/261

1 The Return to Ulster
2 Sweet power of song, duet
3 Once more I hail thee
4 The morning air plays on my face
5 The Massacre of Glencoe
6 What shall I do to shew how much I love her?, duet
7 His boat comes on the sunny tide
8 Come draw we round a cheerful ring
9 The Soldier's Dream
10 The Deserter
11 Thou emblem of faith
12 English Bulls
13 Musing on the roaring ocean
14 Dermot and Shelah
15 Let brain-spinning swains
16 Hide not thy anguish
17 In vain to this desert, duet
18 They bid me slight my Dermot dear, duet
19 Wife, Children and Friends, duet
20 Farewell bliss and farewell Nancy, duet
21 Morning a cruel turmoiler is
22 From Garyone, my happy home
23 A wand'ring gypsy, Sirs, am I
24 The Traugh Welcome
25 Oh harp of Erin

WoO 153 Twenty Irish songs (London and Edinburgh, 1814 [nos.1–4], 1816 [nos.5–10]); GA xxiv/262

1 When eve's last rays, duet
2 No riches from his scanty store
3 The British Light Dragoons
4 Since greybeards inform us
5 I dream'd I lay where flow'rs were springing, duet
6 Sad and luckless was the season
7 O soothe me, my lyre
8 Norah of Balamagairy, with chorus
9 The kiss, dear maid, thy lip has left
10 Oh! thou hapless soldier, duet
11 When far from the home
12 I'll praise the Saints
13 'Tis sunshine at last
14 Paddy O'Rafferty

15 'Tis but in vain
16 O might I but my Patrick love
17 Come, Darby dear
18 No more, my Mary
19 July, lovely, matchless creature
20 Thy ship must sail

WoO 154 Twelve Irish songs (London and Edinburgh, 1816 [without nos.2 and 7]); GA xxiv/258

1 The Elfin Fairies
2 Oh harp of Erin
3 The Farewell Song
4 The pulse of an Irishman
5 Oh! who, my dear Dermot
6 Put round the bright wine
7 From Garyone, my happy home
8 Save me from the grave and wise, with chorus
9 Oh! would I were but that sweet linnet, duet
10 The hero may perish, duet
11 The Soldier in a Foreign Land, duet
12 He promised me at parting, duet

WoO 155 Twenty-six Welsh songs (London and Edinburgh, 1817): GA xxiv/263

1 Sion, the son of Evan, duet
2 The Monks of Bangor's March, duet
3 The Cottage Maid
4 Love without Hope
5 A golden robe my love shall wear
6 The fair Maids of Mona
7 Oh let the night my blushes hide
8 Farewell, thou noisy town
9 To the Aeolian Harp
10 Ned Pugh's Farewell
11 Merch Megan
12 Waken lords and ladies gay
13 Helpless Woman
14 The Dream, duet
15 When mortals all to rest retire
16 The Damsels of Cardigan
17 The Dairy House
18 Sweet Richard

19 The Vale of Clwyd
20 To the Blackbird
21 Cupid's Kindness
22 Constancy, duet
23 The Old Strain
24 Three Hundred Pounds
25 The Parting Kiss
26 Good Night

op.108 Twenty-five Scottish songs (London and Edinburgh, 1818; Berlin, 1822); GA xxiv/257

1 Music, Love, and Wine, with chorus
2 Sunset
3 Oh! sweet were the hours
4 The Maid of Isla
5 The sweetest lad was Jamie
6 Dim, dim is my eye
7 Bonnie laddie, highland laddie
8 The lovely lass of Inverness
9 Behold my love how green the groves, duet
10 Sympathy
11 Oh! thou art the lad
12 Oh, had my fate
13 Come fill, fill, my good fellow, with chorus
14 O, how can I be blithe
15 O cruel was my father
16 Could this ill world
17 O Mary, at thy window be
18 Enchantress, farewell
19 O swiftly glides the bonny boat, with chorus
20 Faithfu' Johnie
21 Jeanie's Distress
22 The Highland Watch, with chorus
23 The Shepherd's Song
24 Again, my lyre
25 Sally in our Alley

WoO 156 Twelve Scottish songs (London and Edinburgh, 1822 [no.1], 1824-5 [nos.2-4, 8, 9, 12], 1839 [nos.5-6], 1841 [nos.7, 10, 11]); GA xxiv/260

1 The Banner of Buccleuch, trio
2 Duncan Gray, trio

3 Up quit thy bower, trio
4 Ye shepherds of this pleasant vale, trio
5 Cease your funning
6 Highland Harry
7 Polly Stewart
8 Womankind, trio
9 Lochnager, trio
10 Glencoe, trio
11 Auld lang syne, trio with chorus
12 The Quaker's Wife, trio

WoO 157 Twelve songs of various nationality (London and Edinburgh, 1816 [nos.2, 6, 8, 11], 1822 [no.3], 1824–5 [no.5], 1839 [no.1]); GA xxiv/259
1 God Save the King (Eng.), with chorus
2 The Soldier (Irish)
3 O Charlie is my darling (Scottish), trio
4 O sanctissima (Sicilian), trio
5 The Miller of the Dee (Eng.), trio
6 A health to the brave (Irish), duet
7 Since all thy vows, false maid (Irish), trio
8 By the side of the Shannon (Irish)
9 Highlander's Lament (Scottish), with chorus
10 Sir Johnie Cope (?Scottish)
11 The Wandering Minstrel (Irish), with chorus
12 La gondoletta (Venetian)

WoO 158a Twenty-three songs of various nationality, *Die Musik*, ii (1902–3) [no.19], J. Schmidt-Görg: *Unbekannte Manuscripte zu Beethovens weltlicher und geistlicher Gesangsmusik* (Bonn, 1928) [no.17], complete (Leipzig, 1943]; HS xiv
1 Ridćer Stig tjener i Congens Gaard (Dan.)
2 Horch auf, mein Liebchen (Ger.)
3 Wegen meiner bleib d'Fräula (Ger.)
4 Warn i in der Früh aufsteh (Tirolean)
5 I bin a Tyroler Bua (Tirolean)
6 A Madel, ja a Madel (Tirolean)
7 Wer solche Buema afipackt (Tirolean)
8 Ih mag di nit (Tirolean)
9 Oj upiłem sie w karczmie (Pol.)
10 Poszła baba po popiół (Pol.)
11 Yo no quiero embarcarme (?Port.)
12 Seus lindos olhos (Port.), duet
13 Im Walde sind viele Mücklein geboren (Russ.)
14 Ach Bächlein, Bächlein, kühle Wasser (Russ.)
15 Unsere Mädchen gingen in den Wald (Russ.)
16 Schöne Minka, ich muss scheiden (Ukrainian: 'Air cosaque')
17 Lilla Carl, sov sött i frid (Swed.)
18 An ä Bergli bin i gesässe (?Swiss)
19 Una paloma blanca (Sp.: 'Bolero a solo')
20 Como la mariposa (Sp.: 'Bolero a due'), duet
21 La tiranna se embarca (Sp.)
22 Édes kinos emlékezet (Hung.)
23 Da brava, Catina (Venetian)

WoO 158b Seven British songs [most texts traced by W. Hess]; HS xiv
1 Adieu my lov'd harp (Irish)
2 Text unidentified (Irish), quartet
3 Oh was not I a weary wight (Scottish)
4 Red gleams the sun (Scottish)
5 Erin! oh, Erin! (Irish or Scottish)
6 O Mary ye's be clad in silk (Scottish)
7 Lament for Owen Roe O'Neill (Irish), text inc.

WoO 158c Six songs of various nationality [most texts traced by Hess]; HS xiv
1 When my hero in court appears (from The Beggar's Opera)
2 Non, non, Collette n'est point trompeuse (from Le devin du village)
3 Mark yonder pomp of costly fashion (Scottish)
4 Bonnie wee thing (Scottish), trio
5 From thee, Eliza, I must go (Scottish), trio
6 Text unidentified

Hess 168 Air français [text unidentified]; HS xiv
— Two Austrian folksongs, with pf acc., *Niederrheinische Musikzeitung*, xiii (1865)
Hess 133 Das liebe Kätzchen
Hess 134 Der Knabe auf dem Berge

Bibliography

CATALOGUES OF WORKS, BIBLIOGRAPHIES, BIBLIOGRAPHICAL STUDIES

Thematisches Verzeichniss der im Druck erschienenen Werke von Ludwig van Beethoven (Leipzig, 1851)

A. W. Thayer: *Chronologisches Verzeichniss der Werke Ludwig van Beethovens* (Berlin, 1865)

G. Nottebohm: *Thematisches Verzeichniss der im Druck erschienenen Werke Ludwig van Beethovens* (Leipzig, 1868)

E. Kastner: *Bibliotheca Beethoveniana* (Leipzig, 1913, enlarged 2/1925 by T. von Frimmel)

'Beethoven-Bibliographie', *NBJb*, i–x (1924–42); *BeJB 1953–*

T. von Frimmel: *Beethoven-Handbuch* (Leipzig, 1926/R1968)

W. Haas: *Systematische Ordnung Beethovenscher Melodien* (Bonn and Leipzig, 1932)

D. W. MacArdle: 'A Check-list of Beethoven's Chamber Music',*ML*, xxvii (1946), 44, 83, 156, 251 [includes substantial bibliography]

A. Bruers: *Beethoven: catalogo storico-critico di tutte le opere* (Rome, 4/1950)

P. Hirsch and C. B. Oldman: 'Contemporary English Editions of Beethoven', *MR*, xiv (1953), 1–35

G. Kinsky and H. Halm: *Das Werk Beethovens: thematisch-bibliographisches Verzeichnis seiner sämtlichen vollendeten Kompositionen* (Munich and Duisburg, 1955) [standard thematic and bibliographical catalogue of Beethoven's works]

P. Nettl: *Beethoven Encyclopedia* (New York, 1956, 2/1967 as *Beethoven Handbook*)

W. Hess: *Verzeichnis der nicht in der Gesamtausgabe veröffentlichten Werke Ludwig van Beethovens* (Wiesbaden, 1957)

A. Tyson: *The Authentic English Editions of Beethoven* (London, 1963)

H. Schmidt: 'Verzeichnis der Skizzen Beethovens', *BeJb 1965–8*, 7–128

G. Biamonti: *Catalogo cronologico e tematico delle opere di Beethoven, comprese quelle inedite e gli abbozzi non utilizzati* (Turin, 1968)

D. W. MacArdle: *Beethoven Abstracts* (Detroit, 1973) [periodical literature]

K. Dorfmüller, ed.: *Beiträge zur Beethoven-Bibliographie: Studien und Materialen zum Werkverzeichnis von Kinsky–Halm* (Munich, 1978) [includes suppl. to Kinsky and Halm]

K. Schürmann, ed.: *Ludwig van Beethoven: alle vertonten und musikalisch bearbeiteten Texte* (Münster, 1980)

193

COLLECTIONS OF ESSAYS AND RELATED PUBLICATIONS

G. Nottebohm: *Beethoveniana* (Leipzig and Winterthur, 1872, 2/1925/*R*1970)

——: *Zweite Beethoveniana*, ed. E. Mandyczewski (Leipzig, 1887, 2/1925/*R*1970)

T. von Frimmel: *Neue Beethoveniana* (Vienna, 1888)

E. Mandyczewski: *Namen- und Sachregister zu Nottebohms Beethoveniana und Zweite Beethoveniana* (Leipzig, 1888/*R*1970)

T. von Frimmel: *Beethoven-Studien* (Munich and Leipzig, 1905–6)

BeethovenJb, ed. T. von Frimmel, i–ii (1908–9)

Beethoven-Forschung, ed. T. von Frimmel, i–iii (1911–25)

A. Sandberger: *Ausgewählte Aufsätze zur Musikgeschichte, ii: Forschungen, Studien und Kritiken zu Beethoven und zur Beethovenliteratur* (Munich, 1924/*R*1970)

Neues Beethoven-Jahrbuch, ed. A. Sandberger, i–x (1924–42) [*NBJb*]

G. Bosse, ed.: *Beethoven-Almanach der Deutschen Musikbücherei auf das Jahr 1927* (Regensburg, 1927)

D. F. Tovey: *Essays in Musical Analysis* (London, 1935–44)

A. Schmitz, ed.: *Beethoven und die Gegenwart: ... Ludwig Schiedermair zum 60. Geburtstag* (Berlin, 1937)

L. Misch: *Beethoven-Studien* (Berlin, 1950; Eng. trans., 1953)

Beethoven-Jahrbuch, ed. J. Schmidt-Görg and others (1953–) [*BeJB*]

Festschrift Joseph Schmidt-Görg zum 60. Geburtstag (Bonn, 1957)

Colloquium amicorum: Joseph Schmidt-Görg zum 70. Geburtstag (Bonn, 1967)

L. Misch: *Neue Beethoven-Studien und andere Themen* (Munich and Duisburg, 1967)

H. C. R. Landon: *Essays on the Viennese Classical Style* (New York, 1970)

E. Schenk, ed.: *Beethoven-Studien: Festgabe der Österreichischen Akademie der Wissenschaften* (Vienna, 1970)

H. Sittner, ed.: *Beethoven-Almanach 1970* (Vienna, 1970)

Internationale Musikwissenschaftliche Kongress: Bonn 1970 (Kassel, 1971)

D. Arnold and N. Fortune, eds.: *The Beethoven Companion* (London, 1971)

P. H. Lang, ed.: *The Creative World of Beethoven* (New York, 1971)

E. Schenk, ed.: *Beethoven-Symposion* (Vienna, 1971)

T. Scherman and L. Biancolli, eds.: *The Beethoven Companion* (New York, 1972)

W. Hess: *Beethoven-Studien* (Munich, 1973)

A. Tyson, ed.: *Beethoven Studies*, i (New York, 1973 and London, 1974); ii (London, 1977); iii (Cambridge, 1982)

Bibliography

H. Goldschmidt: *Beethoven-Studien* (Berlin, 1974–)

Beethoven, Dokumentation und Aufführungspraxis: Vienna 1977, ed. R. Klein (Kassel, 1978)

Internationale Beethoven-Kongress: Berlin 1977, ed. H. Goldschmidt, K.-H. Köhler and K. Niemann (Leipzig, 1978)

H. Goldschmidt, ed.: *Zu Beethoven: Aufsätze und Annotationen* (Berlin, 1979)

C. Wolff, ed.: *The String Quartets of Haydn, Mozart and Beethoven: Studies of the Autograph Manuscripts: Isham Memorial Library 1979*

Beethoven, Performers and Critics: Detroit 1977, ed. R. Winter and B. Carr (Detroit, 1980)

LETTERS, CONVERSATION BOOKS AND OTHER DOCUMENTS

L. Nohl: *Briefe Beethovens* (Stuttgart, 1865)

L. von Köchel, ed.: *Drei und achtzig neu aufgefundene Original-Briefe Ludwig van Beethovens an den Erzherzog Rudolph* (Vienna, 1865)

L. Nohl: *Neue Briefe Beethovens* (Stuttgart, 1867)

——: *Ludwig van Beethovens Brevier* (Leipzig, 1870, 2/1901)

——: *Die Beethoven-Feier und die Kunst der Gegenwart* (Vienna, 1871) [transcr. of Beethoven's diaries, 52ff]

F. Kerst: *Beethoven im eigenen Wort* (Berlin and Leipzig, 1904, 2/1905; Eng. trans., 1905, as *Beethoven, the Man and the Artist, as Revealed in his own Words*)

A. C. Kalischer, ed.: *Beethovens sämtliche Briefe* (Berlin, 1906–8, rev. 2/1909–11 by T. von Frimmel; Eng. trans., 1909)

F. Prelinger, ed.: *Ludwig van Beethoven: sämtliche Briefe und Aufzeichnungen* (Vienna, 1907–11)

L. Schmidt: *Beethoven-Briefe an Nicolaus Simrock, F. G. Wegeler, Eleonore v. Breuning, und Ferd. Ries* (Berlin, 1909)

E. Kastner, ed.: *Ludwig van Beethovens sämtliche Briefe* (Leipzig, 1910, rev., enlarged 2/1923 by J. Kapp) [Kastner–Kapp is standard Ger. edn. of Beethoven's letters]

A. Leitzmann, ed.: *Ludwig van Beethoven: Berichte der Zeitgenossen, Briefe und persönliche Aufzeichnungen* (Leipzig, 2/1921) [transcr. of Beethoven's diaries, ii, 241ff]

M. Unger: *Ludwig van Beethoven und seine Verleger S. A. Steiner und Tobias Haslinger in Wien, Ad. Mart. Schlesinger in Berlin* (Berlin and Vienna, 1921)

W. Nohl: *Ludwig van Beethovens Konversationshefte*, i/1 (Munich, 1924) [8 conversation books of 1819–20]

W. Hitzig: 'Die Briefe Gottfried Christoph Härtels an Beethoven', *ZMw*, ix (1927), 321

O. G. Sonneck: *Beethoven Letters in America* (New York, 1927)

E. H. Müller: 'Beethoven und Simrock', *Simrock-Jb*, ii (1929), 10–62

195

G. Schünemann: *Ludwig van Beethovens Konversationshefte*, i–iii (Berlin, 1941–3) [conversation books from Feb 1818 to July 1823]

J.-G. Prod'homme: *Les cahiers de conversation 1819–1827* (Paris, 1946)

M. Hamburger, ed.: *Beethoven: Letters, Journals and Conversations* (London, 1951)

E. Anderson: 'The Text of Beethoven's Letters', *ML*, xxxiv (1953), 192

D. Weise: *Entwurf einer Denkschrift an das Appelationsgericht* (Bonn, 1953)

——: 'Ungedruckte oder nur teilweise veröffentlichte Briefe Beethovens aus der Sammlung H. C. Bodmer – Zürich', *BeJB 1953–4*, 9–62

S. Ley: *Beethoven: sein Leben in Selbstzeugnissen, Briefen und Berichten* (Vienna, 1954, 2/1970)

D. MacArdle and L. Misch: *New Beethoven Letters* (Norman, Oklahoma, 1957)

J. Schmidt-Görg: *13 unbekannte Briefe an Josephine Gräfin Deym* (Bonn, 1957) [incl. facs. of letters]

E. Anderson, ed. and trans.: *The Letters of Beethoven* (London, 1961)

D. W. MacArdle: *An Index to Beethoven's Conversation Books* (Detroit, 1962)

L. Magnani: *I quaderni di conversazione di Beethoven* (Milan, 1962)

D. von Busch-Weise: 'Beethovens Jugendtagebuch', *SMw*, xxv (1962), 68 [diary 1792–4]

K.-H. Köhler, G. Herre and D. Beck, eds.: *Ludwig van Beethovens Konversationshefte* (Leipzig, 1968–) [i (1972): Feb 1818–March 1820; ii (1976): April–Sept 1820 and June 1822–Feb 1823; iv (1968): Aug–Sept 1823; v (1970): Dec 1823–April 1824; vi (1974): April–Sept 1824; vii (1978): Oct 1824–July 1825; viii (1981): July–Feb 1826; see also D. Beck and G. Herre: 'Anton Schindlers fingierte Eintragungen in den Konversationsheften', *Zu Beethoven*, ed. H. Goldschmidt (Berlin, 1979), 11–89]

M. Braubach: *Die Stammbücher Beethovens und der Babette Koch* (Bonn, 1970)

C. Flamm: 'Ein Verlegerbriefwechsel zur Beethovenzeit', *Beethoven-Studien: Festgabe der Österreichischen Akademie der Wissenschaften*, ed. E. Schenk (Vienna, 1970)

K.-H. Köhler and G. Herre, eds.: *Ludwig van Beethoven: neun ausgewählte Briefe an Anton Schindler* (Leipzig, 1971) [facs. and transcr.]

J. Schmidt-Görg, ed.: *Des Bonner Bäckermeisters Gottfried Fischer Aufzeichnung über Beethovens Jugend* (Bonn and Munich, 1971)

H. Goldschmidt: 'Beethoven in neuen Brunsvik-Briefen', *BeJb 1977*, 97–146

Bibliography

A. Tyson: 'Prolegomena to a Future Edition of Beethoven's Letters', *Beethoven Studies*, ii (London, 1977), 1–32

C. Brenneis: 'Das Fischof-Manuskript: zur Frühgeschichte der Beethoven-Biographik', *Zu Beethoven*, ed. H. Goldschmidt (Berlin, 1979), 90

M. Solomon: 'Beethoven's Tagebuch of 1812–1818', *Beethoven Studies*, iii, ed. A. Tyson (London, 1982), 193–285

BIOGRAPHICAL STUDIES
general

J. A. Schlosser: *Ludwig van Beethoven* (Prague, 1828)

F. G. Wegeler and F. Ries: *Biographische Notizen über Ludwig van Beethoven* (Koblenz, 1838, suppl. Bonn, 1845, both *R*1972; rev. 2/1906 by A. C. Kalischer; Eng. trans., ed. A. Tyson, in preparation)

A. Schindler: *Biographie von Ludwig van Beethoven* (Münster, 1840, enlarged 2/1845, rev. 3/1860; Eng. trans., 1966 as *Beethoven as I Knew him*)

L. Nohl: *Beethovens Leben* (Vienna, 1864 [vol.i]; Leipzig, 1867–77, 2/1906 [completed])

A. W. Thayer: *Ludwig van Beethovens Leben*, trans. H. Deiters, i (Berlin, 1866), ii (Berlin, 1872), iii (Berlin, 1879), vol. 1 rev. H. Deiters (Berlin, 1901); biography continued and completed by H. Deiters and H. Riemann, iv–v (Leipzig, 1907–8); final revisions by H. Riemann, ii–iii (Leipzig, 1910–11), i (Leipzig, 1917); vols. ii–v reissued (Leipzig, 1922–3); Eng. orig., ed. and rev. H. Krehbiel (New York, 1921), rev. E. Forbes as *Thayer's Life of Beethoven* (Princeton, 1964, 2/1967)

T. von Frimmel: *Ludwig van Beethoven* (Berlin, 1901, 6/1922)

R. Rolland: *La vie de Beethoven* (Paris, 1907; numerous later edns.)

F. Kerst, ed.: *Die Erinnerungen an Beethoven* (Stuttgart, 1913, 2/1925)

E. Ludwig: *Beethoven: Life of a Conqueror* (New York, 1943)

W. Hess: *Beethoven* (Zurich, 1956, enlarged 2/1976)

G. Marek: *Beethoven: Biography of a Genius* (New York, 1969)

iconographies, picture biographies

T. von Frimmel: *Beethoven-Studien*, i: *Beethovens äussere Erscheinung* (Munich, 1905)

——: *Beethoven im zeitgenössischen Bildnis* (Vienna, 1923)

S. Ley: *Beethovens Leben in authentischen Bildern und Texten* (Berlin, 1925, enlarged 2/1970)

R. Bory: *Ludwig van Beethoven: his Life and his Work in Pictures* (Zurich and New York, 1960)

J. Schmidt-Görg and H. Schmidt, eds.: *Ludwig van Beethoven* (Bonn, 1969; Eng. trans., 1970)

H. C. R. Landon, ed.: *Beethoven: a Documentary Study* (London and New York, 1970) *character and personality*

A. de Hevesy: *Beethoven: vie intime* (Paris, 1926; Eng. trans., 1927, as *Beethoven the Man*)

G. Adler: *Beethovens Charakter* (Regensburg, 1927)

E. Newman: *The Unconscious Beethoven* (London, 1927, rev. 2/1969)

M. Solomon: 'Beethoven's Birth Year', *MQ*, lvi (1970), 702

——: 'Beethoven: the Nobility Pretense', *MQ*, lxi (1975), 272

——: 'The Dreams of Beethoven', *American Imago*, xxxii (1975), 113–44

——: 'Beethoven and his Nephew: a Reappraisal', *Beethoven Studies*, ii, ed. A. Tyson (London, 1977), 138

education

I. von Seyfried, ed.: *Ludwig van Beethovens Studien im Generalbass, Contrapunkt und in der Compositionslehre* (Vienna, 1832, 2/1853/*R*1967; Eng. trans., 1853)

G. Nottebohm: 'Generalbass und Compositionslehre betreffende Handschriften Beethovens', *Beethoveniana* (Leipzig and Winterthur, 1872, 2/1925/*R*1970), 154–203

——: *Beethovens Studien* (Leipzig and Winterthur, 1873/*R*1970) [general refutation of Seyfried (1832)]

A. Orel, ed.: *Ein Wiener Beethovenbuch* (Vienna, 1921)

G. Schünemann: 'Beethovens Studien zur Instrumentation', *NBJb*, viii (1938), 146

A. Mann: 'Beethoven's Contrapuntal Studies with Haydn', *MQ*, lvi (1970), 711

R. Kramer: 'Beethoven and Carl Heinrich Graun', *Beethoven Studies*, i, ed. A. Tyson (New York, 1973), 18

——: 'Notes to Beethoven's Education', *JAMS*, xxviii (1975), 72–101

Beethoven and his contemporaries

G. von Breuning: *Aus dem Schwarzspanierhause: Erinnerungen an L. van Beethoven aus meiner Jugendzeit* (Leipzig and Vienna, 1874, rev. 2/1907 by A. C. Kalischer)

L. Nohl: *Eine stille Liebe zu Beethoven* (Leipzig, 1875, 2/1901; Eng. trans., 1876) [diaries of Fanny Giannatasio del Rio]

A. C. Kalischer: *Beethoven und seine Zeitgenossen* (Berlin, 1908–10)

O. G. Sonneck, ed.: *Beethoven: Impressions of Contemporaries* (New York, 1926/*R*1967)

S. Ley: *Beethoven als Freund der Familie Wegeler–von Breuning* (Bonn, 1927) [incl. part of F. G. Wegeler and F. Ries: *Biographische Notizen über Ludwig van Beethoven*, and G. Bruening: *Aus dem Schwarzspanierhause*]

J. Heer: *Der Graf von Waldstein und sein Verhältnis zu Beethoven* (Leipzig, 1933)

Bibliography

D. W. MacArdle: 'Beethoven and George Thomson', *ML*, xxxvii (1956), 27

———: 'Beethoven and the Bach Family', *ML*, xxxix (1958), 353

———: 'Beethoven and the Czernys', *MMR*, lxxxviii (1958), 124

———: 'Beethoven and Grillparzer', *ML*, xl (1959), 44

———: 'Beethoven and Haydn', *MMR*, lxxxix (1959), 203

———: 'Beethoven and the Archduke Rudolph', *BeJb 1959–60*, 36

———: 'Anton Felix Schindler, Friend of Beethoven', *MR*, xxiv (1960), 50

———: 'Beethoven and Ferdinand Ries', *ML*, xlvi (1965), 23

———: 'Beethoven and Schuppanzigh', *MR*, xxvi (1965), 3

———: 'Beethoven und Karl Holz', *Mf*, xx (1967), 19

M. Solomon: 'Beethoven and Bonaparte', *MR*, xxix (1968), 96

———: 'Antonie Brentano and Beethoven', *ML*, lviii (1977)

———: 'Schubert and Beethoven', *19th Century Music*, iii (1979–80), 114

———: 'Beethoven and Schiller', *Beethoven, Performers, and Critics*, ed. R. Winter and B. Carr (Detroit, 1980), 162

the 'Immortal Beloved'

M. Tenger: *Beethovens unsterbliche Geliebte nach persönlichen Erinnerungen* (Bonn, 1890; Eng. trans., 1898, as *Recollections of Countess Therese Brunswick*)

La Mara [pseud. of M. Lipsius]: *Beethovens unsterbliche Geliebte* (Leipzig, 1909)

W. A. Thomas-San-Galli: *Die unsterbliche Geliebte Beethovens, Amalie Sebald* (Halle, 1909)

M. Unger: *Auf Spuren von Beethovens unsterblicher Geliebten* (Langensalza, 1911)

La Mara [pseud. of M. Lipsius]: *Beethoven und die Brunsviks* (Leipzig, 1920)

O. G. Sonneck: *The Riddle of the Immortal Beloved* (New York, 1927)

S. Kaznelson: *Beethovens ferne und unsterbliche Geliebte* (Zurich, 1954)

M. Solomon: 'New Light on Beethoven's Letter to an Unknown Woman', *MQ*, lviii (1972), 572 [generally acknowledged as giving the correct identification: Antonie Brentano]

H. Goldschmidt: *Beethoven-Studien*, ii: *Um die unsterbliche Geliebte: ein Bestandsaufnahme* (Leipzig, 1977)

M. Solomon: 'Antonie Brentano and Beethoven', *ML*, lviii (1977), 153

special studies

W. Schweisheimer: *Beethovens Leiden: ihr Einfluss auf sein Leben und Schaffen* (Leipzig, 1922)

J. M. Levien: *Beethoven and the Royal Philharmonic Society* (London, 1927)

W. Nohl: *Beethoven: Geschichten und Anekdoten* (Berlin, 1927)

R. Van Aerde: *Les ancêtres flamands de Beethoven* (Malines, 1927)

E. Closson: *L'élément flamand dans Beethoven* (Brussels, 1928, 2/1946; Eng. trans., 1936)

J. Boyer: *Le 'romantisme' de Beethoven* (Paris, 1939)

D. W. MacArdle: 'The Family van Beethoven', *MQ*, xxxv (1949), 528 [incl. substantial bibliography]

B. Bartels: *Beethoven und Bonn* (Dinkelsbühl, 1954)

E. and R. Sterba: *Beethoven and his Nephew* (New York, 1954)

W. Forster: *Beethovens Krankheiten und ihre Beurteilung* (Wiesbaden, 1955)

S. Ley: *Aus Beethovens Erdentagen* (Siegburg, 1957)

D. W. MacArdle: 'Beethoven and the Philharmonic Society of London', *MR*, xxi (1960), 1

L. G. Bachmann: *Beethoven contra Beethoven: Geschichte eines berühmten Rechtfalles* (Munich, 1963)

J. Schmidt-Görg: *Beethoven: die Geschichte seiner Familie* (Munich, 1964) [the standard genealogy]

A. Tyson: 'Beethoven's Heroic Phase', *MT*, cx (1969), 139

R. Klein: *Beethovenstätten in Österreich* (Vienna, 1970)

E. Larkin: 'Beethoven's Medical History', suppl. chap. to M. Cooper: *Beethoven: the Last Decade* (London, 1970), 439

K. Smolle: *Wohnstätten Ludwig van Beethovens von 1792 bis zu seinem Tod* (Munich and Duisburg, 1970)

F. Knight: *Beethoven and the Age of Revolution* (London, 1973)

LIFE-AND-WORK STUDIES

W. von Lenz: *Beethoven: eine Kunst-Studie* (Kassel and Hamburg, 1855–60, rev. 2/1908 by A. C. Kalischer)

A. B. Marx: *Ludwig van Beethoven: Leben und Schaffen* (Berlin, 1859, 6/1911)

W. J. von Wasielewski: *Ludwig van Beethoven* (Berlin, 1888, 2/1895)

G. Grove: 'Beethoven, Ludwig van', *Grove 1*; repr. in *Beethoven–Schubert–Mendelssohn* (London, 1951)

P. Bekker: *Beethoven* (Berlin and Leipzig, 1911, numerous later edns.; Eng. trans., 1925)

V. d'Indy: *Beethoven: biographie critique* (Paris, 1911, 2/1927; Eng. trans., 1912)

W. A. Thomas-San-Galli: *Ludwig van Beethoven* (Munich, 1913, 8/1920)

J. G. Prod'homme: *La jeunesse de Beethoven* (Paris, 1920, 3/1937)

L. Schiedermair: *Der junge Beethoven* (Leipzig, 1925, 3/1951)

Bibliography

H. Grace: *Ludwig van Beethoven* (London, 1927)

A. Schmitz: *Beethoven* (Bonn, 1927)

R. Rolland: *Les grands époques créatrices* (Paris, 1928–57; Eng. trans. of vol. i, 1929, as *Beethoven the Creator*; vol.ii, 1931, as *Goethe and Beethoven*)

R. H. Schauffler: *Beethoven: the Man who Freed Music* (London and New York, 1929)

F. Howes: *Beethoven* (London, 1933)

E. Bücken: *Ludwig van Beethoven* (Potsdam, 1934)

M. Scott: *Beethoven* (London, 1934, rev. 2/1974 by J. A. Westrup)

W. Riezler: *Beethoven* (Berlin and Zurich, 1936, 9/1966; Eng. trans., 1938/R1972)

A. E. F. Dickinson: *Beethoven* (London, 1941)

J. N. Burk: *The Life and Works of Beethoven* (New York, 1943)

J. Schmidt-Görg: 'Beethoven, Ludwig van', *MGG*

M. Cooper: *Beethoven: the Last Decade, 1817–1827* (London, 1970)

M. Solomon: *Beethoven* (New York, 1977)

STUDIES OF THE WORKS
general and historical assessments

W. von Lenz: *Beethoven et ses trois styles* (St Petersburg, 1852–3, 2/1909)

A. Oulibicheff: *Beethoven, ses critiques et ses glossateurs* (Leipzig and Paris, 1857)

H. Berlioz: *A travers chants* (Paris, 1862; Eng. trans., 1913–18); ed. L. Guichard (Paris, 1971); extracts ed. and trans. R. De Sola as *Beethoven by Berlioz* (Boston, Mass., 1975)

R. Wagner: *Beethoven* (Leipzig, 1870, numerous later edns.; Eng. trans., 1880)

D. G. Mason: *Beethoven and his Forerunners* (New York, 1904, 2/1930)

F. Volbach: *Beethoven* (Munich, 1905)

E. Walker: *Beethoven* (London, 1905, 3/1920)

G. Becking: *Studien zu Beethovens Personalstil: das Scherzothema* (Leipzig, 1921)

H. Mersmann: *Beethoven: die Synthese der Stile* (Berlin, 1922)

A. Schmitz: *Beethovens 'zwei Prinzipe'* (Berlin and Bonn, 1923)

F. Cassirer: *Beethoven und die Gestalt: ein Kommentar* (Stuttgart, 1925)

A. Halm: *Beethoven* (Berlin, 1926/R1971)

K. Kobald: *Beethoven: seine Beziehungen zu Wiens Kunst und Kultur* (Vienna and elsewhere, 1926, numerous later edns.)

A. Schmitz: *Das romantische Beethoven-Bild* (Berlin and Bonn, 1927)

J. W. N. Sullivan: *Beethoven: his Spiritual Development* (London, 1927, 2/1936)

Beethoven

D. F. Tovey: 'Some Aspects of Beethoven's Art-forms', *ML*, viii (1927), 131; repr. in *Essays and Lectures on Music* (London, 1949)

W. J. Turner: *Beethoven: the Search for Reality* (London, 1927, 2/1933)

A. Schering: *Beethoven in neuer Deutung* (Leipzig, 1934)

D. F. Tovey: *Beethoven* (London, 1944)

K. von Fischer: *Die Beziehung von Form und Motiv in Beethovens Instrumentalwerken* (Strasbourg and Zurich, 1948, enlarged 2/1972)

L. Ronga: *Bach, Mozart, Beethoven: tre problemi critici* (Venice, 1956)

L. Misch: *Die Faktoren der Einheit in der Mehrsätzigkeit der Werke Beethovens* (Munich and Duisburg, 1958)

C. Rosen: *The Classical Style* (New York, 1971)

M. Solomon: 'The Creative Periods of Beethoven', *MR*, xxxiv (1973), 30

H. Goldschmidt: *Beethoven: Werkeinführungen* (Leipzig, 1975)

——: 'Beethoven in der Analyse', *BMw*, xix/1 (1977)

C. Rosen: *Sonata Forms* (New York, 1980)

D. Johnson: '1794–1795: Decisive Years in Beethoven's Early Development', *Beethoven Studies*, iii, ed. A. Tyson (London, 1982), 1

J. Kerman: 'Notes on Beethoven's Codas', *Beethoven Studies*, iii, ed. A. Tyson (London, 1982), 141

orchestral music

G. Grove: *Beethoven's Nine Symphonies* (London, 1884, enlarged 2/1896 as *Beethoven and his Nine Symphonies*, 3/1898/R1962)

J.-G. Prod'homme: *Les symphonies de Beethoven* (Paris, 1906, 5/1949)

F. Weingartner: *Ratschläge für Aufführungen der Symphonien Beethovens* (Leipzig, 1906, rev. 2/1916; Eng. trans., 1907/R in *Weingartner on Music and Conducting*, New York, 1969)

H. Schenker: *Beethovens Neunte Sinfonie* (Vienna and Leipzig, 1912/R1969)

——: *Beethovens Fünfte Sinfonie* (Vienna, 1925/R1969; partial Eng. trans. in *Beethoven: Symphony no.5 in C minor*, ed. E. Forbes, New York, 1971)

J. Braunstein: *Beethovens Leonore-Ouvertüren* (Leipzig, 1927)

K. Nef: *Die neun Sinfonien Beethovens* (Leipzig, 1928/R1970)

D. F. Tovey: *Beethoven's Ninth Symphony* (London, 1928)

A. Halm: 'Über den Wert musikalischer Analysen, i: Der Fremdkörper im ersten Satz der Eroica', *Die Musik*, xxi/2 (1929), 48

H. Schenker: 'Beethovens Dritte Sinfonie, in ihrem wahren Inhalt zum erstenmal dargestellt', *Das Meisterwerk in der Musik*, iii (Munich, 1930/R1974), 29–101

R. Vaughan Williams: *Some Thoughts on Beethoven's Choral Symphony, with other Musical Subjects* (London, 1953)

W. Osthoff: *Ludwig van Beethoven: Klavierkonzert Nr. 3, C-moll op.37* (Munich, 1965)

Bibliography

C. Westphal: *Vom Einfall zur Symphonie: Einblick in Beethovens Schaffensweise* (Berlin, 1965) [on Sym. no.2]

A. Tyson: 'The Textual Problems of Beethoven's Violin Concerto', *MQ*, liii (1967), 482

F. E. Kirby: 'Beethoven's Pastoral Symphony as a *Sinfonia caracteristica*', *MQ*, lvi (1970), 605

L. Lockwood: 'Beethoven's Unfinished Piano Concerto of 1815: Sources and Problems', *MQ*, lvi (1970), 624

E. Forbes, ed.: *L. van Beethoven: Symphony no.5 in C minor* (New York, 1971) [Norton Critical Score]

F. d'Amico: *Sulle sinfonie di Beethoven* (Rome, 1973)

P. Gossett: 'Beethoven's Sixth Symphony: Sketches for the First Movement', *JAMS*, xxvii (1974), 248–84

T. Antonicek: 'Humanitätssymbolik im Eroica-Finale', *De ratione in musica: Festschrift Erich Schenk* (Kassel, 1975), 144

A. Tyson: 'The Problem of Beethoven's "First" *Leonore* Overture', *JAMS*, xxviii (1975), 292–334

——: 'Yet Another "Leonore" Overture?', *ML*, lviii (1977), 192

P. Gülke: *Zur Neuausgabe der Sinfonie Nr.5 von Ludwig van Beethoven: Werk und Edition* (Leipzig, 1978)

L. Treitler: 'History, Criticism, and Beethoven's Ninth Symphony', *19th Century Music*, iii (1979–80), 193

A. Hopkins: *The Nine Symphonies of Beethoven* (London and Seattle, 1980)

R. Winter: 'The Sketches for the "Ode to Joy"', *Beethoven, Performers, and Critics*, ed. R. Winter and B. Carr (Detroit, 1980), 176–214

L. Lockwood: 'Beethoven's Earliest Sketches for the *Eroica* Symphony', *MQ*, lxvii (1981), 467

——: 'Eroica Perspectives: Strategy and Design in the First Movement', *Beethoven Studies*, iii, ed. A. Tyson (Cambridge, 1982), 85

chamber music

T. Helm: *Beethovens Streichquartette: Versuch einer technischen Analyse* (Leipzig, 1885, 3/1921)

H. J. Wedig: *Beethovens Streichquartett op.18, 1 und seine erste Fassung* (Bonn, 1922)

J. de Marliave: *Les quatuors de Beethoven* (Paris, 1925; Eng. trans., 1928/R1961)

M. Herwegh: *Technique et interprétation sous forme d'essai d'analyse psychologique expérimentale appliquée aux sonates pour piano et violon de Beethoven* (Paris, 1926)

C. Engel: 'Beethoven's Opus 3 – an "envoi de Vienne"?', *MQ*, xiii (1927), 261

W. Engelsmann: *Beethovens Kompositionspläne, dargestellt in den Sonaten für Klavier und Violine* (Augsburg, 1931)

D. G. Mason: *The Quartets of Beethoven* (New York, 1947)

C. B. Oldman: 'Beethoven's "Variations on National Themes": their Composition and First Publication', *MR*, xii (1951), 45

D. Cooke: 'The Unity of Beethoven's Late Quartets', *MR*, xxiv (1963), 30

W. Kirkendale: 'The "Great Fugue" Op.133: Beethoven's "Art of Fugue" ', *AcM*, xxxv (1963), 14

I. Mahaim: *Beethoven: naissance et renaissance des derniers quatuors* (Paris, 1964)

P. Radcliffe: *Beethoven's String Quartets* (London, 1965/*R*1978)

W. Kirkendale: *Fuge und Fugato in der Kammermusik des Rokoko und der Klassik* (Tutzing, 1966; rev. and Eng. trans., 2/1979)

J. Kerman: *The Beethoven Quartets* (New York, 1967)

H. Truscott: *Beethoven's Late String Quartets* (London, 1968)

E. Krefft: *Die späten Quartette Beethovens* (Bonn, 1969)

D. Johnson: 'Beethoven's Sketches for the Scherzo of the Quartet Op.18, No.6', *JAMS*, xxiii (1970), 385

L. Lockwood: 'The Autograph of the First Movement of Beethoven's Sonata for Violoncello and Pianoforte, Opus 69', *Music Forum*, ii (1970), 1–109

R. Stephan: 'Zu Beethovens letzten Quartetten', *Mf*, xxiii (1970), 245

A. Tyson: 'Stages in the Composition of Beethoven's Piano Trio op.70 no.1', *PRMA*, xcvii (1970–71), 1

J. D. Kramer: 'Multiple and Non-linear Time in Beethoven's Opus 135', *PNM*, xi/2 (1973), 122; see also J. Lochhead: 'The Temporal in Beethoven's *Opus 135*', *In Theory Only*, liv (Jan 1979), 3

M. Mila: 'Lettura della "Grande fuga" op.133', *Scritti in onore di Luigi Ronga* (Milan and Naples, 1973), 345

W. J. Mitchell: 'Beethoven's La Malinconia from the String Quartet, Opus 18, No.6: Technique and Structure', *Music Forum*, iii (1973), 269

G. Boringhieri: 'Le due sonate di Beethoven, Op.102, n.1 e 2, per pianoforte e violoncello', *NRMI*, xi (1977), 537

S. Brandenburg: 'Bemerkungen zu Beethovens op.96', *BeJb 1977*, 11

——: 'The First Version of Beethoven's G major String Quartet, Op.18 No.2', *ML*, lviii (1977), 127

R. Winter: 'Plans for the Structure of the String Quartet in C sharp minor', *Beethoven Studies*, ii, ed. A. Tyson (London, 1977), 106

F. Eibner: 'Einige Kriterien für die Apperzeption und Interpretation von Beethovens Werk', *Beethoven-Kolloquium 1977*, ed. R. Klein (Kassel, 1978), 20 [analysis of op.24]

L. Lockwood: 'Beethoven's Early Works for Violoncello and Contemporary Violoncello Technique', *Beethoven-Kolloquium 1977*, ed. R. Klein (Kassel, 1978), 174

S. Brandenburg: 'The Autograph of Beethoven's String Quartet in A minor, Opus 132', *The Quartets of Haydn, Mozart, and Beethoven*, ed. C. Wolff (Cambridge, Mass., 1980), 278 .

R. Kramer: ' "Das Organische der Fuge": On the Autograph of Beethoven's String Quartet in F major, Opus 59, No.1', *The Quartets of Haydn, Mozart, and Beethoven*, ed. C. Wolff (Cambridge, Mass., 1980), 223–65

M. Staehelin: 'Another Approach to the Last String Quartet Oeuvre: the Unfinished String Quintet of 1926/27', *The Quartets of Haydn, Mozart, and Beethoven*, ed. C. Wolff (Cambridge, Mass., 1980), 302

J. Webster: 'Traditional Elements in Beethoven's Middle-Period String Quartets', *Beethoven, Performers, and Critics*, ed. R. Winter and B. Carr (Detroit, 1980), 94–133

A. Glauert: 'The Double Perspective in Beethoven's Opus 131', *19th Century Music*, iv (1980–1), 113

S. Brandenburg: 'The Historical Background to the "Heiliger Dank-gesang" in Beethoven's A-minor Quartet, op.132', *Beethoven Studies*, iii, ed. A. Tyson (Cambridge, 1982), 161–91

R. Kramer: 'Ambiguities in *La Malinconia*: What the Sketches Say', *Beethoven Studies*, iii, ed. A. Tyson (Cambridge, 1982), 29

A. Tyson: 'The "Razumovsky" Quartets: Some Aspects of the Sources', *Beethoven Studies*, iii, ed. A. Tyson (Cambridge, 1982), 107–40

piano music

A. B. Marx: *Anleitung zum Vortrag Beethovens Klavierwerke* (Berlin, 1863, 2/1875)

C. Reinecke: *Die Beethovenschen Clavier-Sonaten* (Leipzig, 1895)

W. Nagel: *Beethoven und seine Klaviersonaten* (Langensalza, 1903–5, 2/1923–4)

H. Schenker: *Die letzten Sonaten von Beethoven: kritische Ausgabe mit Einführung und Erläuterung* (Vienna, 1913–21, rev., abridged 2/1971–2 by O. Jonas); see review by W. Drabkin, *PNM*, xii (1973–4), 319

H. Riemann: *L. van Beethovens sämtliche Klavier-Solosonaten* (Berlin, 1918–19, 4/1920)

D. F. Tovey: *A Companion to Beethoven's Pianoforte Sonatas* (London, 1931)

J.-G. Prod'homme: *Les sonates pour piano de Beethoven (1782–1823)* (Paris, 1937, 2/1950)

E. Blom: *Beethoven's Pianoforte Sonatas Discussed* (London, 1938/R1968)

E. Hertzmann: 'The Newly Discovered Autograph of Beethoven's Rondo a capriccio Op.129', *MQ*, xxxii (1946), 171

E. Fischer: *Ludwig van Beethoven's Klaviersonaten* (Wiesbaden, 1956, 2/1966; Eng. trans., 1959)

R. Rosenberg: *Die Klaviersonaten Ludwig van Beethovens* (Olten, 1957)

J. V. Cockshoot: *The Fugue in Beethoven's Piano Music* (London, 1959)

A. Forte: *The Compositional Matrix* (Baldwin, NY, 1961) [on op.109]

C. Czerny: *Über den richtigen Vortrag der sämtlichen Beethoven'schen Klavierwerke*, ed. P. Badura-Skoda (Vienna, 1963; Eng. trans., 1970) [orig. part of suppl. to Czerny's *Complete Theoretical and Practical Pianoforte School* op.500]

R. Réti: *Thematic Patterns in the Sonatas of Beethoven* (London, 1965)

W. S. Newman: 'The Performance of Beethoven's Trills', *Beethoven-Jahrbuch,* ix (1973–7), 347–76, also pubd in *JAMS,* xxix (1976),

H. Federhofer: 'Zur Analyse des zweiten Satzes von L. van Beethovens Klaviersonate op.10, Nr.3', *Festskrift Jens Peter Larsen* (Copenhagen, 1972), 339

W. S. Newman: 'The Performance of Beethoven's Trills', *Beethoven-Jahrbuch,* ix (1973–7), 347–76, also pubd in *JAMS*, xxix (1976), 439: see also replies by R. Winter and W. S. Newman in *MQ*, lxiii (1977), 483; lxiv (1978), 98; lxv (1979), 111

A. Leicher-Olbrich: *Untersuchungen zu Originalausgaben Beethovenscher Klavierwerke* (Wiesbaden, 1976)

E. Cone: 'Beethoven's Experiments in Composition: the Late Bagatelles', *Beethoven Studies*, ii, ed. A. Tyson (London, 1977), 84

C. Timbrell: 'Notes on the Sources of Beethoven's Opus 111', *ML*, lviii (1977), 204

R. Kramer: 'On the Dating of Two Aspects in Beethoven's Notation for Piano', *Beethoven-Kolloquium 1977*, ed. R. Klein (Kassel, 1978), 160

W. Drabkin: 'Beethoven's Sketches and the Thematic Process', *PRMA*, cv (1978–9), 25 [on op.111]

C. Reynolds: 'Beethoven's Sketches for the Variations in E♭, Op.35', *Beethoven Studies*, iii, ed. A. Tyson (Cambridge, 1982), 47

vocal music

G. Schünemann: 'Beethovens Skizzen zur Kantate "Der glorreiche Augenblick" ', *Die Musik*, ix (1909–10), 22, 93

M. Kufferath: *Fidelio de L. van Beethoven* (Paris, 1913)

H. Boettcher: *Beethoven als Liederkomponist* (Augsburg, 1928)

F. Lederer: *Beethovens Bearbeitungen schottischer und anderer Volkslieder* (Bonn, 1934)

C. Hopkinson and C. B. Oldman: 'Thomson's Collections of National Song, with Special Reference to the Contributions of Haydn and Beethoven', *Transactions of the Edinburgh Bibliographical Society*,

Bibliography

ii, pt.i (1940); addenda et corrigenda, iii, pt.ii (1954), 123

W. Hess: *Beethovens Oper Fidelio und ihre drei Fassungen* (Zurich, 1953)

T. Adorno: 'Verfremdetes Hauptwerk: zur Missa solemnis', *Prisma der gegenwärtigen Musik*, ed. J. E. Berendt and J. Uhde (Heidelberg, 1959); repr. in *Moments musicaux* (Frankfurt am Main, 1964)

E. Anderson: 'Beethoven's Operatic Plans', *PRMA*, lxxxviii (1961–2), 61

W. Hess: *Beethovens Bühnenwerke* (Göttingen, 1962)

E. Forbes: 'Sturzet nieder Millionen', *Studies in Music History: Essays for Oliver Strunk* (Princeton, 1968), 449

J. Lester: 'Revisions in the Autograph of the *Missa Solemnis Kyrie*', *JAMS*, xxiii (1970), 420

A. Tyson: 'The 1803 Version of Beethoven's Christus am Oelberge', *MQ*, lvi (1970), 551–84

J. Kerman: 'An die ferne Geliebte', *Beethoven Studies*, i, ed. A. Tyson (New York, 1973), 123–57

M. Ruhnke: 'Die Librettisten des *Fidelio*', *Opernstudien: Anna Amalie Abert zum 65. Geburtstag* (Tutzing, 1975), 121

S. Howell: 'The Maelzel Canon: another Schindler Forgery?', *MT*, cxx (1979), 987

CATALOGUES AND DESCRIPTIONS OF BEETHOVEN ARCHIVES

G. Adler: *Verzeichniss der musikalischen Autographe von Ludwig van Beethoven . . . im Besitze von A. Artaria in Wien* (Vienna, 1890)

J. S. Shedlock: 'Beethoven's Sketch Books', *MT*, xxxiii (1892), 331, 394, 461, 523, 589, 649; xxxiv (1893), 14, 53; xxxv (1894), 13, 449, 596 [MSS in *GB-Lbm*]

A. Artaria: *Verzeichnis von musikalischen Autographen . . . vornemlich der reichen Bestände aus dem Nachlasse . . . Ludwig van Beethoven's* (Vienna, 1893)

A. C. Kalischer: 'Die Beethoven-Autographe der Königlichen Bibliothek zu Berlin', *MMg*, xxvii (1895), 145; xxviii (1896), 1–80

A. Hughes-Hughes: *Catalogue of Manuscript Music in the British Museum* (London, 1906–9/R1964–6)

G. Kinsky: *Musikhistorisches Museum von Wilhelm Heyer in Cöln: Katalog*, iv (Cologne, 1916)

J. Schmidt-Görg: *Katalog der Handschriften des Beethoven-Hauses und Beethoven-Archivs* (Bonn, 1935)

M. Unger: 'Die Beethovenhandschriften der Pariser Konservatoriumsbibliothek', *NBJb*, vi (1935), 87–123

——: 'Die Beethovenhandschriften der Familie W. in Wien', *NBJb*, vii (1937), 155 [Wittgenstein collection]

——: *Eine Schweizer Beethovensammlung* (Zurich, 1939) [collection of H. C. Bodmer]

G. Kinsky: *Manuskripte, Briefe, Dokumente von Scarlatti bis Stravinsky: Katalog der Musikautographensammlung Louis Koch* (Stuttgart, 1953)

N. Fischmann [Fishman]: 'Autographen Beethovens in der UdSSR', *BMw*, iii/1 (1961), 22

B. Schwarz: 'Beethoveniana in Soviet Russia', *MQ*, xlvii (1961), 4 [see also *MQ*, xlix (1963), 143]

J. Kerman: 'Beethoven Sketchbooks in the British Museum', *PRMA*, xciii (1966–7), 77

H. Schmidt: 'Die Beethovenhandschriften des Beethovenhauses in Bonn', *BeJb 1969–70*, pp.vii–xxiv, 1–443; suppl., *BeJb 1971–2*, 207

O. E. Albrecht, H. Cahoon and D. Ewing, eds.: *The Mary Flagler Cary Music Collection* (New York, 1970)

E. Bartlitz: *Die Beethoven-Sammlung in der Musikabteilung der Deutschen Staatsbibliothek: Verzeichnis* (Berlin, 1970)

P. J. Willetts: *Beethoven and England: an Account of the Sources in the British Museum* (London, 1970)

D. Johnson: 'The Artaria Collection of Beethoven Manuscripts: a New Source', *Beethoven Studies*, i, ed. A. Tyson (New York, 1973)

H.-G. Klein: *Ludwig van Beethoven: Autographe und Abschriften*, Staatsbibliothek Preussischer Kulturbesitz: Kataloge der Musikabteilung, i/2 (Berlin, 1975)

EDITIONS OF AUTOGRAPHS AND SKETCHBOOKS
facsimile editions of major autographs

Piano Sonata, A♭, op.26: ed. E. Prieger (Bonn, 1895)

Piano Sonata, c♯, op.27 no.2: ed. H. Schenker (Vienna, 1927) [with facs. of 3 sketch leaves]; new edn. by K. Sakka (Tokyo, 1970) [with facs. of 1st edn. of sonata]

Violin Sonata, G, op.30 no.3: ed. A. Tyson (London, 1980)

Piano Sonata, C, op.53: ed. D. Weise (Bonn, 1954)

Piano Sonata, f, op.57: (Paris, c1926/Rc1970)

String Quartet, F, op.59 no.1: ed. A. Tyson (London, 1980)

String Quartet, e, op.59 no.2: ed. A. Tyson (London, 1980)

Symphony no.5, op.67: ed. G. Schünemann (Berlin, 1942)

Cello Sonata, A, op.69: ed. L. Lockwood (New York, 1970) [1st movt]

Piano Sonata, F♯, op.78 (Munich, 1923)

Violin Sonata, G, op.96 (Munich, 1976)

An die ferne Geliebte, op.98 (Munich, 1970)

Piano Sonata, E, op.109: ed. O. Jonas (New York, 1965)

Piano Sonata, A♭, op.110: ed. K.-M. Komma (Stuttgart, 1967) [with transcr. of corrections in autograph and of relevant sketches]

Bibliography

Piano Sonata, c, op.111 (Munich, 1922/*Rc*1952 and 1969)
Missa solemnis, op.123: ed. W. Virneisel (Tutzing, 1965) [1st movt]
Symphony no.9, d, op.125 (Leipzig, 1924/*R*1975)

sketchbooks in facsimile, or in complete or substantial transcription
G. Nottebohm: *Ein Skizzenbuch von Beethoven* (Leipzig, 1865, 2/1924/*R*1970) [sketchbook of 1801–2: substantial transcr.]
——: *Ein Skizzenbuch von Beethoven aus dem Jahre 1803* (Leipzig, 1880, 2/1924/*R*1970) [substantial transcr.]
C. de Roda: 'Un quaderni di autografi del 1825', *RMI*, xii (1905), 63–108, 592–622, 732–67; pubd separately (Turin, 1907) [substantial transcr.]
Beethovens eigenhändiges Skizzenbuch zur 9. Symphonie, ed. W. Engelmann (Leipzig, 1913) [facs.]
M. Iwanow-Boretzky: *Ein Moskauer Skizzenbuch von Beethoven* (Vienna, 1927; first pubd in *Muzïkal'noe obrazovanie* (1927), Jan–March 9–58 [facs.], 75–91 [commentary]) [sketches for String Quartets opp.130 and 132]
K. L. Mikulicz: *Ein Notierungsbuch von Beethoven aus dem Besitz der Preussischen Staatsbibliothek zu Berlin* (Leipzig, 1927/*R*) [transcr.; sketchbook of 1800–01]
J. Schmidt-Görg: *Beethoven: Drei Skizzenbücher zur Missa Solemnis* (Bonn, 1952–70) [facs. and transcr.]
D. Weise: *Beethoven: Ein Skizzenbuch zur Chorfantasie op.80 und zu anderen Werken* (Bonn, 1961) [transcr.]
——: *Beethoven: Ein Skizzenbuch zur Pastoralsymphonie op.68 und zu den Trios op.70* (Bonn, 1961) [transcr.]; see review by L. Lockwood, *MQ*, liii (1967), 128
N. Fishman: *Kniga eskizov Beethoven za 1802–1803 gody* (Moscow, 1962) [facs. and transcr.]
J. Schmidt-Görg: *Beethoven: Ein Skizzenbuch zu den Diabelli-Variationen und zur Missa Solemnis, SV 154* (Bonn, 1968–72) [facs. and transcr.]; see review by R. Winter, *JAMS*, xxviii (1975), 135
J. Kerman: *Ludwig van Beethoven: Autograph Miscellany from circa 1786 to 1799* (London, 1970) [facs. and transcr.]
W. Virneisel: *Beethoven: Ein Skizzenbuch aus Streichquartetten aus op.18, SV 46* (Bonn, 1972–4) [facs. and transcr.]
S. Brandenburg: *Beethoven: Kesslersches Skizzenbuch* (Bonn, 1976–8) [facs. and transcr.; facs. also pubd separately (Munich, 1976)]
D. Johnson: *Beethoven's Early Sketches in the 'Fischhof Miscellany'* (Ann Arbor, 1980) [transcr.]

OTHER STUDIES

M. Unger: *Beethoven über eine Gesamtausgabe seiner Werke* (Bonn, 1920)

P. Mies: *Die Bedeutung der Skizzen Beethovens zur Erkenntnis seines Stiles* (Leipzig, 1925; Eng. trans., 1929/*R*1969, 1974)

M. Unger: *Beethovens Handschrift* (Bonn, 1926)

E. Newman: *The Unconscious Beethoven* (London, 1927, rev. 2/1969)

T. Veidl: *Der musikalische Humor bei Beethoven* (Leipzig, 1929)

E. Brümmer: *Beethoven im Spiegel der zeitgenössischen Presse* (Würzburg, 1932)

G. Kinsky: 'Zur Versteigerung von Beethovens musikalischem Nachlass', *NBJb*, vi (1935), 66

A. Schering: *Beethoven und die Dichtung* (Berlin, 1936/*R*1973)

L. Schrade: *Beethoven in France* (New Haven, 1942)

P. Mies: *Textkritische Untersuchungen bei Beethoven* (Munich and Duisburg, 1957)

H. Unverricht: *Die Eigenschriften und Originalausgaben von Werken Beethovens in ihrer Bedeutung für die moderne Textkritik* (Kassel, 1960)

A. Tyson: 'Maurice Schlesinger as a Publisher of Beethoven', *AcM*, xxxv (1963), 182

H. Grundmann and P. Mies: *Studien zum Klavierspiel Beethovens und seiner Zeitgenossen* (Bonn, 1966, 2/1970)

P. Stadlen: 'Beethoven and the Metronome: I', *ML*, xlviii (1967), 330

J. Kerman: 'Beethoven's Early Sketches', *MQ*, lvi (1970), 515

L. Lockwood: 'On Beethoven's Sketches and Autographs: Some Problems of Definition and Interpretation', *AcM*, xlii (1970), 32

A. Tyson: 'Notes on Five of Beethoven's Copyists', *JAMS*, xxiii (1970), 439–71

M. Solomon: 'Beethoven, Sonata, and Utopia', *Telos* (1971), no.9, p.32

A. Tyson: 'Sketches and Autographs'; 'Steps to Publication – and Beyond', *The Beethoven Companion*, ed. D. Arnold and N. Fortune (London, 1971), 443; 459

D. Johnson and A. Tyson: 'Reconstructing Beethoven's Sketchbooks', *JAMS*, xxv (1972), 137

A. Tyson: 'A Reconstruction of the Pastoral Symphony Sketchbook', *Beethoven Studies*, i (1973), 67

M. Solomon: 'Beethoven and the Enlightenment', *Telos* (1974), no.19, p.146

A. Tyson: 'Das Leonoreskizzenbuch (Mendelssohn 15): Probleme der Rekonstruktion und der Chronologie', *BeJb 1977*, 469

R. W. Wade: 'Beethoven's Eroica Sketchbook', *FAM*, xxiv (1977), 254–89

Bibliography

D. Johnson: 'Beethoven Scholars and Beethoven's Sketches', *19th Century Music*, ii (1978–9), 3; see also S. Brandenburg, W. Drabkin and D. Johnson: 'On Beethoven Scholars and Beethoven's Sketches', *ibid*, iii (1979–80), 270

S. Brandenburg: 'Ein Skizzenbuch Beethovens aus dem Jahre 1812: Zur Chronologie des Petterschen Skizzenbuches', *Zu Beethoven*, ed. H. Goldschmidt (Berlin, 1979), 117–48

M. Solomon: 'On Beethoven's Creative Process: a Two-part Invention', *ML*, lxi (1980), 272

Index

Albrechtsberger, Johann Georg, 14, 15, 49, 125, 134
Amenda, Karl, 26, 30, 31, 100, 139
Anschütz, Heinrich, 88
Artaria, 20, 82, 146
Aschaffenburg, 9
Augsburg, 5

Bach, Carl Philipp Emanuel, 49
Bach, Johann Baptist, 67, 68
Bach, Johann Sebastian, 3, 16, 125, 130, 155
Baden, 39, 44, 51, 53, 81, 83, 84, 143
Bartók, Béla, 151
Beethoven, Caspar Anton Carl van [Beethoven's brother], 2, 21, 33, 34, 35, 49, 57, 63
Beethoven (née Reiss), Johanna van [Beethoven's sister-in-law], 33, 63, 64, 65, 66, 67, 68, 71, 83, 84, 85
Beethoven, Johann van [Beethoven's father], 1, 2, 5, 12, 13
Beethoven, Karl van [Beethoven's nephew], 33, 57, 63, 64, 65, 66, 67, 68, 70, 83, 84, 85, 86, 88, 141
Beethoven, Ludwig van [Beethoven's grandfather], 1
Beethoven, Ludwig Maria van [Beethoven's brother], 2
Beethoven (née Poll), Maria Josepha van [Beethoven's grandmother], 1
Beethoven (née Keverich), Maria Magdalena van [Beethoven's mother], 1, 5
Beethoven, Johann van [Beethoven's brother], 2, 21, 34, 53, 54, 85

Beethoven (née Obermeyer), Therese van [Beethoven's sister-in-law], 54, 85
Bellini, Vincenzo, 121
Berlin, 21, 22, 23, 74, 78, 83, 156
——, Singakademie, 22
Berlioz, Hector, 121, 150, 155
Bernadotte, 23
Blöchlinger, Joseph, 67, 83
Boehm, Joseph, 81
Bonaparte, Jerome, 47
Bonaparte, Napoleon: *see* Napoleon
Bonn, 1, 3, 5, 6, 7, 9, 10, 11, 12, 13, 14, 16, 17, 18, 19, 21, 22, 29, 30, 32, 51, 75, 89, 92, 94, 95, 99, 121, 140, 141, 144, 146, 156
——, Beethovenhaus, 152, 153
——, Bonngasse, 2
——, Lesegesellschaft, 7
——, Zehrgarten, 10
Bouilly, J. N., 38
Brahms, Johannes, 151, 154
Bratislava: *see* Pressburg
Braun, Baron, 41
Breitkopf & Härtel, 35, 53, 146
Brentano, Antonie, 55, 56, 70
Brentano, Bettina, 51, 55, 151, 152
Brentano, Franz, 55, 56
Brentano, Maximiliane, 55, 56
Breuning, Christoph von, 9, 22
Breuning, Eleonore von, 18, 19, 22
Breuning, Gerhard, 88
Breuning, Lorenz von, 9, 22
Breuning, Stephan von, 9, 39, 85, 139
Bridgetower, George Polgreen, 37
Browne, Count Johann Georg von, 23
Bruckner, Anton, 150

212

Index

213

Index

Index

215

216